Lecture Notes on Data Engineering and Communications Technologies

Volume 40

Series Editor

Fatos Xhafa, Technical University of Catalonia, Barcelona, Spain

The aim of the book series is to present cutting edge engineering approaches to data technologies and communications. It will publish latest advances on the engineering task of building and deploying distributed, scalable and reliable data infrastructures and communication systems.

The series will have a prominent applied focus on data technologies and communications with aim to promote the bridging from fundamental research on data science and networking to data engineering and communications that lead to industry products, business knowledge and standardisation.

** Indexing: The books of this series are submitted to ISI Proceedings, MetaPress, Springerlink and DBLP **

More information about this series at http://www.springer.com/series/15362

Aneta Poniszewska-Marańda · Natalia Kryvinska ·
Stanisław Jarząbek · Lech Madeyski

Editors

Data-Centric Business
and Applications

Towards Software Development (Volume 4)

 Springer

Editors
Aneta Poniszewska-Marańda
Institute of Information Technology
Lodz University of Technology
Łódź, Poland

Stanisław Jarząbek
Faculty of Computer Science
Bialystok University of Technology
Białystok, Poland

Natalia Kryvinska
Department of e-Business, Faculty
of Business, Economics and Statistics
University of Vienna
Vienna, Wien, Austria

Lech Madeyski
Faculty of Computer Science
and Management
Wrocław University of Science
and Technology
Wrocław, Poland

ISSN 2367-4512   ISSN 2367-4520   (electronic)
Lecture Notes on Data Engineering and Communications Technologies
ISBN 978-3-030-34705-5   ISBN 978-3-030-34706-2   (eBook)
https://doi.org/10.1007/978-3-030-34706-2

This Springer imprint is published by the registered company Springer Nature Switzerland AG
The registered company address is: Gewerbestrasse 11, 6330 Cham, Switzerland

Preface

With the fourth volume we continue to analyze challenges and opportunities for doing business with information emphasizing development of the software from different points of view. We cover also methods and techniques, as well as strategies for the efficient software production toward business information processing and management.

Explicitly, starting with the first chapter "Towards a Unified Requirements Model for Distributed High Performance Computing", the authors discuss the conceptual model, functional scope, and research agenda for creating and validating a Distributed High-Performance Computing (DHPC) software development system that fulfills defined requirements. A generic requirements model consisting of a conceptual domain specification, unified domain vocabulary, and use-case-based functional requirements is proposed. Vocabulary definition provides detailed clarifications of HPC fundamental component elements and their role in the system. Moreover, the authors describe the security issues by providing transparency principles for HPC. A research agenda that leads to the creation of a model-based software development system dedicated to building Distributed HPC applications at a high level of abstraction is also proposed.

In the next chapter titled "Requirement Engineering as a Software Development Process", definition of model of requirements management platform aimed at Agile practitioners to bridge the gap between source code tooling and requirements engineering is given. The chapter describes a requirements management tool, incorporating software development practices into requirements engineering. The authors' aim is to provide an open architecture for various requirements engineering activities by proposing a mapping of various aspects of software development based on Continuous Integration to requirements engineering and describing a prototype of requirements management tool built to validate the proposed concept.

In the work called "Information Management System for an Administrative Office with RFID-Tagged Classified Documents", the authors present the development of system for electronic and paper document traceability based on RFID tags. Starting from describing the system architecture for the RFID-equipped restricted access administrative office, the chapter deals with hardware and software components as well as business and simulation models which are the result of analytical work performed by the group of experts. Such group was composed of specialist in different fields, such as document management, IT, data security, and radio frequency identification to develop and evaluate the system proposition based on computer simulation.

The chapter authored by Bogumiła Hnatkowska and Martyna Litkowska, "Framework for Processing Behavioral Business Rules Written in a Controlled Natural Language" presents the idea of processing the behavioral business rules that were written in controlled natural language that is highly recommended to be understandable for all interested parties. The paper presents the state of a framework for business rules processing to be able to serve the business rules written in controlled language. As the authors mentioned, there exists a very limited number of solutions enabling processing business rules expressed that way. The proof-of-concept implementation that proved the correctness and usefulness of the proposed approach was also presented.

"Software Defect Prediction Using Bad Code Smells: A Systematic Literature Review" presents the state of the art in the field of fault prediction models that include code smell information—the authors give the current state of the art in the field of bug prediction with the use of code smells and attempt to identify the areas requiring further research. To achieve this goal, a systematic literature review of 27 research papers published between 2006 and 2019 was conducted. For each paper, the reported relationship between smelliness and bugginess was analyzed, as well as the performance of code smell data used as a defect predictor in models developed using machine learning techniques was evaluated. The results of this investigation confirm that code smells are positively correlated with software defects and can positively influence the performance of fault detection models.

The chapter "Software Development Artifacts in Large Agile Organizations: A Comparison of Scaling Agile Methods" focuses on higher complexity related to multiple value streams development pipelines orchestration, communications among many distributed teams, inter-team dependencies management, and information flow between teams. The purpose of the authors is to compare the possible agile frameworks for scaling development organizations working in an agile culture, and their outcomes materialized as artifacts.

The work entitled "Tabu Search Algorithm for Vehicle Routing Problem with Time Windows" emphases the transportation as an important task in the society because economies of modern world are based on internal and foreign trade. The author focuses on one of the problems in the field of transportation which is Vehicle Routing Problem (VRP) proposing the solution to some aspects of it using the general vehicle routing heuristics needed for real-life problems. One of them is Tabu Search that is presented as an efficient algorithm to solve Vehicle Routing Problem with Time Windows constraint. The optimistic and interesting results of using the algorithms for benchmark cases were also presented. Moreover, the obtained results were compared with world's best values to show that the implementation of heuristic has improved the best known solutions to benchmark cases for many problems.

In the chapter authored by Hanna Grodzicka, Arkadiusz Ziobrowski, Zofia Łakomiak, Michał Kawa, and Lech Madeyski, "Code Smell Prediction Employing Machine Learning Meets Emerging Java Language Constructs" contributes to definition of code smell that recognition tends to be highly subjective. However, there exist some code smells detection tools and some of them use the machine learning techniques to overcome the disadvantages of automatic detection tools. The main purpose of the authors was to develop a research infrastructure and reproduce the process of code smell prediction proposed by Arcelli Fontana et al. To do that they investigated machine learning algorithms performance for samples including major modern Java language features. This study was performed with dataset of 281 Java projects. The detection rules derived from the best performing algorithms incorporated newly introduced metrics that were described.

The next chapter "Cloud Cognitive Services Based on Machine Learning Methods in Architecture of Modern Knowledge Management Solutions" presents the concept of cloud cognitive services as cloud computing services available to help developers to build the intelligent applications based on Machine Learning methods with pretrained models as a service. Currently, Machine Learning platform is one of the fastest growing services of the cloud because machine learning and artificial intelligence platforms are available through diverse delivery models such as cognitive computing, automated machine learning, and model management. The author proposes a new cognitive service based approach to build an architecture of knowledge management system. The possibilities of using cognitive service were analyzed and some of the relevant aspects of cloud cognitive service and machine learning in knowledge management context were discussed.

In the next chapter titled "Anti-Cheat Tool for Detecting Unauthorized User Interference in the Unity Engine Using Blockchain", the concept of blockchain was used for security aspects such as detecting of unauthorized user interference in the Unity engine. The main aim of the authors was to analyze the problem of cheating in online games and to design a comprehensive tool for the Unity engine that detects and protects the applications against unauthorized interference by a user. To do that, the basic functions such as detecting and blocking of unauthorized interference in the device's memory, detecting the modification of the speed of time flowing the game, and detecting time changes in the operating system were introduced. The analysis of current potential of blockchain technology as a secure database for the game was also presented.

In the chapter "Approaches to Business Analysis in Scrum at StepStone—Case Study", the author discusses the possibilities of using the agile methods for different areas of a business analysis because the opinions about the role of a Business Analyst in Agile teams vary greatly—from negation of its existence to acceptance of a business analyst as a team member, working side by side with programmers. The author presents his own experience is this domain, experimented with several models of team organization, with or without dedicated BAs. Two most prominent cases are presented: when the BA was a full-time member of the development team and when the BA's role was distributed between other team members—developers and Product Owner. Moreover, the advantages, disadvantages, and the transition process are discussed together with the techniques to help in development of necessary analytical skills and transforming team organization.

The chapter called "Development Method of Innovative Projects in Higher Education Based on Traditional Software Building Process" refines the earlier established metrics of project quality and project efficiency, categorizes them along the proposed success dimensions, providing the necessary adaptations for an academic setting and generalizing them so that they can be applied to a broad spectrum of student software undertakings. The authors focus on the students' projects and their success metrics such as three dimensions of success that have been elicited basing on prior industrial studies: project quality, project efficiency as well as social factors (teamwork quality and learning outcomes).

A further chapter "Light-Weight Congestion Control for the DCCP: Implementation in the Linux Kernel" aims to present the prototype implementation of the light-weight DCCP's congestion control algorithm, designed for multimedia transmission. This algorithm was based on the RTP linear throughput equation and the prototype implementation in the Linux kernel includes a new congestion control module and updates to the DCCP kernel API. The implementation was tested and the results of the tests were presented by the authors.

And, the final chapter "Light-Weight Congestion Control for the DCCP: Implementation in the Linux Kernel" presents the prototype implementation of modified TFRC congestion control, designed for multimedia transmission. The described implementation in Linux kernel includes both a new congestion control module and updates to the DCCP kernel API. The proposed solution causes the DCCP to be not fully TCP-friendly, but still remains TCP-tolerant and does not cause unnecessary degradation of competing TCP flows.

Łódź, Poland   Aneta Poniszewska-Marańda
                                                     aneta.poniszewska-maranda@p.lodz.pl
Vienna, Austria   Natalia Kryvinska
                                                       natalia.kryvinska@univie.ac.at
Białystok, Poland   Stanisław Jarząbek
                                                             s.jarzabek@pb.edu.pl
Wrocław, Poland   Lech Madeyski
                                                           lech.madeyski@pwr.edu.pl

Contents

Towards a Unified Requirements Model for Distributed High Performance Computing

Michał Śmiałek, Kamil Rybiński, Radosław Roszczyk and Krzysztof Marek

Abstract High Performance Computing (HPC) consists in development and execution of sophisticated computation applications, developed by highly skilled IT personnel. Several past studies report significant problems with applying HPC in industry practice. This is caused by lack of necessary IT skills in developing highly parallelised and distributed computation software. This calls for new methods to reduce software development effort when constructing new computation applications. In this paper we propose a generic requirements model consisting of a conceptual domain specification, unified domain vocabulary and use-case-based functional requirements. Vocabulary definition provides detailed clarifications of HPC fundamental component elements and their role in the system. Further we address security issues by providing transparency principles for HPC. We also propose a research agenda that leads to the creation of a model-based software development system dedicated to building Distributed HPC applications at a high level of abstraction, with the object of making HPC more available for smaller institutions.

1 Introduction and Motivation

One of the critical problems in technology innovation activities is the efficient processing of large amounts of data using sophisticated computation algorithms. Computations like this need highly efficient supercomputing systems which are expensive, complex to use and can be configured and programmed only by highly skilled IT personnel. Such requirements pose a big problem for smaller groups like small and medium companies (SMEs), innovative start-ups or small research institutions. These organisations are unable to make a significant financial commitment. Despite having

M. Śmiałek (✉) · K. Rybiński · R. Roszczyk · K. Marek
Warsaw University of Technology, Warsaw, Poland
e-mail: michal.smialek@ee.pw.edu.pl

K. Rybiński
e-mail: kamil.rybinski@ee.pw.edu.pl

© Springer Nature Switzerland AG 2020
A. Poniszewska-Marańda et al. (eds.), *Data-Centric Business and Applications*,
Lecture Notes on Data Engineering and Communications Technologies 40,
https://doi.org/10.1007/978-3-030-34706-2_1

necessary domain knowledge (e.g. solving engineering problems), due to their lack of software development competencies, they are unable to benefit from HPC.

Furthermore, High-Performance Computing (HPC) [12] is also in the process of dynamic changes. This change is associated with the emergence of Cloud Computing and its application to HPC [30]. This new trend can be compared to older paradigms like Grid Computing, and generally—to distributed systems [9]. While growth in hardware and low-level operating system capabilities is tremendous, the capabilities to develop HPC software seem to lag behind. As Giles and Reguly [10] point out, this is caused by quite slow adaptation of existing code libraries to new architectures and parallel distributed processing. Moreover, orchestration of computations in a distributed, cloud-enabled environment poses significant problems to programmers [25]. Considering this, Van De Vanter et al. [29] observe the need for new software development infrastructures, explicitly dedicated to HPC. An interesting study by Schmidberger and Brugge [26] shows that understanding of software engineering methods among HPC application developers is not yet satisfactory. What is more, these methods need to be significantly adapted to the specificity of scientific and engineering computations. An example approach to this issue is presented by Li et al. [17, 18]. It proposes a lightweight framework to define requirements for computation applications using a domain-specific approach [5].

Currently, many companies, who are willing to take a leap into HPC solutions, are forced to use outside organizations to rewrite their original programs. For them, it is the only way to make their solution parallelised and optimised. This process often takes months or even years, because of a lack of domain knowledge among HPC specialists that rewrite code. A good example is the SHAPE project, whose main goal is to provide SMEs with both HPC experts and computation power. Despite such huge help, the results are only able to work on specific architectures and require further development [28]. Moreover, in many cases, the projects were discontinued after the end of SHAPE support. They often did not translate into broader use of HPC in these SMEs. This shows how difficult it is for smaller organisations to use HPC, despite significant benefits.

Therefore we can argue that raising the level of abstraction when programming an HPC application can significantly reduce complexity of programs and increase programmer productivity. In more traditional approaches, this leads to constructing code libraries optimised for HPC, compatible with more modern—object-oriented— programming languages (e.g. Java as compared to FORTRAN or C) [22]. Such approaches allow for better modularisation of code which becomes more elegant, easy to understand and portable. However, they do not remove technological complexity associated with programming platforms and constructs of any contemporary programming language. Even the solutions whose main purpose was to enable easy communication management in object-oriented programming languages, turn out to require complex knowledge of HPC. For example, OpusJava [15, 16] was designed to enable high performance computing in large distributed systems by hiding many low-level details, including thread synchronization and communication. As a result, it made parallel programming easier for IT specialists. Unfortunately it didn't remove the need for sophisticated HPC knowledge and still necessitates significant programming skills. We argue that in order to cope with such complexity we should use model-driven techniques.

Considerations on model-driven approaches lead to the construction of certain domain-specific languages oriented specifically on developing HPC applications. An important problem when developing such languages is to understand the sophisticated nature of large-scale scientific and engineering computations. This is reflected by the complexity and variety of computation problems, structure of the computation algorithms and distribution of computation tasks, as pointed out in past research [2, 6, 23]. Moreover, it can be noted that the terminology for HPC is not standardised. For instance, Dongarra et al. argue about the meaning of certain terms related to clustering and parallelisation [8].

In this position paper we propose an initial conceptual model, functional scope and research agenda for creating and validating a Distributed HPC (DHPC) software development system that fulfils the above general requirements. The goal is to offer a coherent holistic solution: an easy-to-program and affordable supercomputing platform that could improve innovative potential and reduce time-to-market for high-technology services and products. At the same time, it would allow to develop a gradually developing network of independent computation resources that would follow the same underlying computation model and language.

2 Related Work

The idea to use high level of abstraction as a solution to the problem of accessibility to HPC by SMEs and small research institutions, is not new. The Legion system [11] is one of the first attempts to provide functional requirements for distributed HPC. In this work, DHPC is referred-to as the Worldwide Virtual Computer. Much have changed since that moment but the need for unified functional requirements remains. More recently, such requirements can be expressed using domain specific languages. For instance, the Neptune system [6] formulates a formally defined language for specifying the distribution aspects for HPC applications. Other approaches concentrate on specific application domains. For instance, ExaSlang [27] was developed to specify numerical solvers based on the multigrid method. Another example is the system developed by Hernandez et al. [13] to automate the development of grid-based applications. However, we can argue that an ultimate solution should be independent of a specific problem domain or class of computation problems. Such a goal was the basis for constructing HPCML (HPC Modelling Language) [24]. The language uses graphical notation to denote flow of computation tasks, parallelism of tasks and data, and other issues. However, it concentrates only on the computation aspect, and does not describe variuos other aspects necessary to organise the whole DHPC environment.

As far as model transformation techniques are concerned, Palyart et al. propose an interesting method called MDE4HPC [23] that is based on developing higher-level computation models that can be transformed to code. These models abstract away all the platform-specific issues and concentrate on the essence—the computation problem. Another approach is proposed by Membarth et al. [20]. It involves a domain-specific language which is used to develop computation models. The underlying transformation engine can generate code for various processing environments (CPUs and GPUs) thus handling the important problem of heterogeneity in HPC.

Analysis of existing literature shows very limited research that includes approaches to formulate conceptual models and functional requirements for DHPC systems. To our best knowledge, supported by queries to ACM DL, IEEE Xplore, Scopus and Google Scholar, there are no direct solutions of this kind. Analysis of literature shows that research focuses mostly on solving specific problems associated with HPC. Authors do not specify any functional requirements for their solutions, nor any conceptual models. They provide non-functional requirements (e.g. performance parameters), describe their technical solution and assume implementation of certain user expectations that are not explicitly formulated as functional requirements. A good example of such practice is the XtremWeb project [7]. Its goal was to create a peer-to-peer computation system, that would enable parallel computing in a distributed large-scale system. Available literature on XtremWeb focuses on the technical solution, system capabilities and security aspects. User requirements are briefly mentioned, and no conceptual model is given. In our opinion, one of the reasons behind lack of universality in current solutions is different understanding of user expectations. Therefore unifying the domain model and functional requirements for DHPC systems, could ensure broader and more universal solutions in the future.

If we summarise the above considerations we can note that the ultimate goal is to create an integrated general-purpose software development system dedicated to HPC applications. Such system should allow for developing high-level computation models through a domain-specific language that abstracts away the platform-specific issues and concentrates on the essence—defining the computation itself. At the same time, the system should be able to translate this DSL-defined computation into a well-architected computation application. This application should follow best design practices for HPC [1] and be able to quickly react to changing requirements [19]. Finally, it should scale to cover variable models of computing resource utilisation [30].

In addition to many technical aspects that should be taken into account to build a distributed computing system, security of the solution should be taken into account. One of the most important aspects is trust management in a distributed system [3]. The security of a distributed system is generally more complex than the security of an independent system. Current computer security concepts assume that trust is assigned to an element of a distributed system based on a point of view. This security mechanism for distributed file systems solves many performance and security problems in current systems [4, 14].

3 Domain Model for Distributed HPC

In order to obtain a unified domain model for HPC we consider four fundamental elements of any computation problem: the Computation Application, the Computation Task, the Computation Module and the Computation Resource. These four elements are highlighted in Fig. 1. Generally, HPC computations are defined through Computation Applications (CA) that constitute high-level computation "programs". These programs can be provided by end users of a computational platform or by advanced developers involved in the creation of such applications. The CAs are composed of calls to self-contained Computation Modules (CM). Each CM performs a well-defined, atomic piece of calculations. Calculations at this level can be performed independently of each other in a parallel or a sequential way. Execution (instantiation) of a CA and the contained CMs is performed through creating a Computation Task (CT). Each task can be executed on several Computation Resources (CR), i.e. on specific machines (computer hardware with OS) equipped with specialized computing nodes (virtual machines or docker containers).

Each Computation Application represents a sequence of CA Steps (see again Fig. 1) needed to perform the given computation (the algorithm). Within the algorithm, we can distinguish two types of calculation steps. The first type (User Interaction) carries out all kinds of interactions with the users. The second type (Module Call) is responsible for calling specific Computation Modules and controlling of data flow between the modules. These two types of steps are ordered in sequences, which compose into a specific calculation algorithm that realizes a specific end user request. Depending on the construction of the algorithm, some steps may need to be performed in parallel, while others must be performed sequentially. When computations are performed in parallel, a significant issue is to synchronise data produced by the modules working in parallel. This needs to be done before going to the next steps that normally would require data produced by the previous parallel modules.

It can be noted that the conceptual model in Fig. 1 does not present any details regarding the flow of CA Steps. This should be part of a detailed meta-model for a new Computation Application Language which is out of scope of this paper. This language would define constructs for representing flow of computations and data in a distributed computing environment, such as:

- algorithm branching for parallel execution,
- synchronization of data from parallel steps,
- retrying tasks in case of failure (in accordance with the defined retry policy),
- data transfer between computation modules,
- parallelization of computation modules,
- data exchange between modules,
- execution of computation modules.

It would also include constructs to define details of user interactions, including such operations as:

- entering parameters for the algorithms contained in the computation modules,

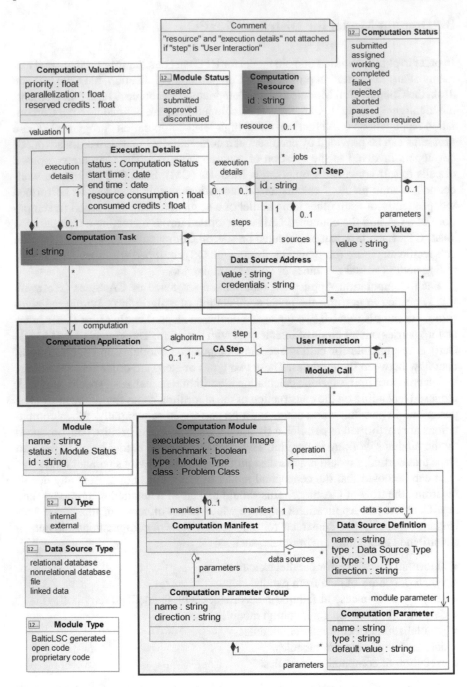

Fig. 1 Domain model for distributed HPC computations

- determining input sources for computation modules,
- determining pointers to storage for the calculation results,
- entering parameters that control computations (performance, security etc.).

Let us now consider the role of Computation Modules. Their primary function is to store and the executable code for a given computational component in the form of a Container Image. The size of this image depends on the actual computation algorithm. The module contents can range from simple algebraic operations to complex algorithms that implement AI tasks. For every CM we have to determine its type and class. The Module Type determines the level of access to the executable module code, and thus determines the level of trust during its execution. This mechanism has been introduced to secure the computing platform against unauthorized use. It also protects the interests of participants offering their machines for computational purposes. The Problem Class applies to the mechanisms associated with benchmarking of COmputation Resources, as explained further in this section.

Each Computation Module, apart from the executable code, is equipped with a Computation Manifest. It aggregates information about data sources (Data Source Definition) and possible computation parameters of the calculation module itself (like computation performance) and input parameters for the computation algorithm. This information thus provides complete information about the input and the output for a given module and assures a mechanism of cooperation between modules. In addition, this manifest contains information on the possibility of parallelisation of the computation task and requirements for possible synchronization between the modules.

The underlying general assumption is that each CM processes some data on its input and generates some data on its output. The computations are controlled by the module parameters (e.g. maximum number of iterations, calculation threshold, cooperation with the environment, data exchange, and other parameters). We should also note the assumption that data is normally not transferred directly to the CM's storage space. Instead, each CM accesses the data through a specific separator (proxy) that gives access to the external or internal storage of the data.

Each execution of a Computation Application with the related CMs, creates a new instance of Computation Task. This task is described by so-called execution details (see the Execution Details class in Fig. 1). They contain performance parameters, such as the consumption of computing resources in the form of CPU usage time, number of cores, memory occupancy, disk space usage, the amount of data transmitted by the network and other parameters of the execution environment. Before starting any computation task, the end-user must declare an estimated target resource consumption for a given CT. This initial estimate can be based on the requirements of the computational algorithm (provided by the algorithm developer) and the amount of data that will be processed. At the stage of task execution planning, the system will take into account the parameters declared by the user and—based on them—appropriately arrange the CT task to be carried out.

When creating a performance plan for a CT, the computation runtime system creates an appropriate CT Step for each CA Step contained in the CA's algorithm.

A unique status is created for each CT Step (Computation Status). The mechanism for collecting and supervising statuses allows for ongoing analysis of the progress of calculations in a distributed environment. In addition, each CT Step stores specific values for its data sources and the current module execution parameters. The actual time to determine these values depends on the sequence of CA Steps and can also be determined by the user interactions (details are out of scope of this paper). The data collected by the computation runtime engine is needed to create future execution plans. In addition, the data stored in each CT can be used in the cases when the computation module has to be stopped in order to transfer it to another computing location (e.g. when the given Computation Resource has to be shut down).

As already indicated, Computation Tasks are executed through running their steps on a network of distributed Computation Resources. The appropriate model is presented in Fig. 2. Each Computation Resource can run several jobs defined through CT Steps (compare also with Fig. 1) Depending on the actual arrangement of data sources and parallelisation, jobs coming from a single Computation Task can be assigned to the same or different Computation Resources.

Similarly to Computation Modules, also CRs have their manifests (CR Manifest). The primary purpose is to collect and store information about the individual hardware configuration and available software within the computing node. These parameters can specify a single machine as well as a group of machines constituting a computational cluster. Within the manifest, virtual launch environments for containers (cf. Docker [21]) can also be defined. This solution significantly simplifies the management of the computing network. All configurations are assembled in a central catalogue (CR Catalogue) and contain descriptions of individual nodes of the calculation network. Each entry defines specific processor parameters or processors available for calculations, graphics processor parameters, available memory, disk resources, and essential operating system parameters within which the jobs assigned to a CR can be executed.

Apart from the manifest for a single CR, the CR owner can also define information on time slots during which computational tasks can be carried out. This can be defined for certain CR Groups. We can also notice that the manifest mechanism allows for collecting and managing information about the actual performance of computing nodes through benchmarking (see Performance Measure in Fig. 2). Performance benchmarking is realized by executing specific calculation modules that serve as actual reference points for specific classes of computational problems (cf. Problem Class). Besides this mechanism, node performance can be calculated based on historical records of already performed calculation modules. What is important, the CR Manifest must be matched with the computational manifests attached to each CM. In this way, the computing network can determine if the CR meets the minimum requirements to run a specific CM. The information contained in manifests and the data on the declared time availability of nodes and information from performance tests are used by the mechanism that optimizes plans made for computational algorithms. Based on these plans, calculations can be assigned to the appropriate nodes.

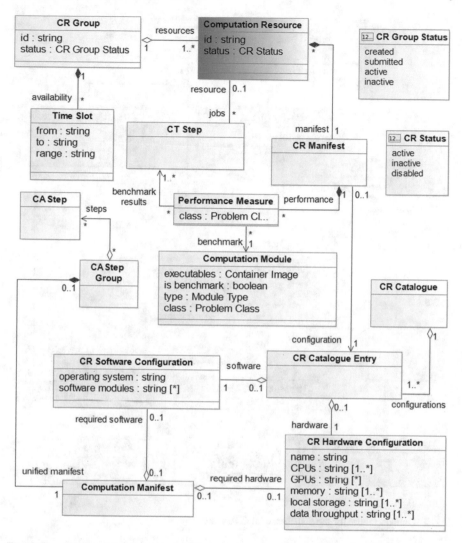

Fig. 2 Domain model for distributed HPC resources

The above computation-related elements have to be structured for storage and for presentation to the various actors of an DHCP system. Each element receives an appropriate metric containing basic data related to its purpose. This leads us to Fig. 3 that presents such structuring elements. The main element that organises Computation Applications or its element is the App Store. It contains all the approved and available CAs. In order for an application to be used by an end-user or by an app-developer, it needs to be selected and added to the App Shelf, associated with as Account with proper credentials. All the Computation Tasks within an Account are organised into a Computation Rack. From the presentation point of view, the two

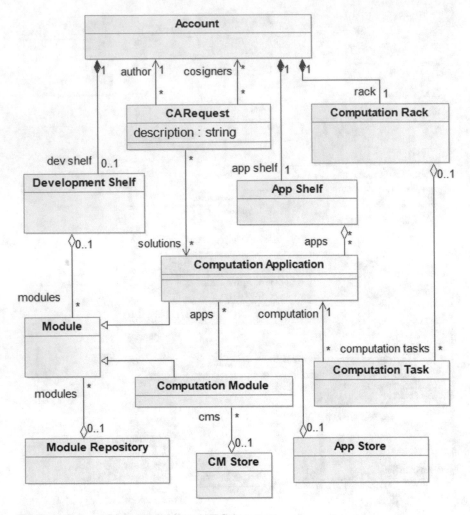

Fig. 3 Domain model for the distributed HPC App store

organising elements (App Shelf and Computation Rack) are shown in the form of a so-called Computation Cockpit.

In addition, for the end-user, who is the developer of computational applications, the Computation Cockpit provides access to the programming shelf. This shelf, in addition to ready-made calculation modules or their elements, contains the basic elements of the system in the form of shares or stocks. This mechanism is used to group all modules (CM and CA) developed in a given account. The application creator may also have unfinished modules over which development work is currently underway. All modules completed and sent by the developers are stored in the Module Repository. Each time a module is accepted, it is added to the App Store (CA) or the CM Store (CM). As part of the development account, apart from creating new

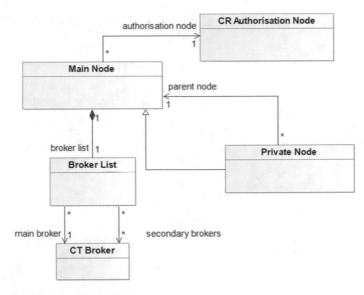

Fig. 4 Domain model for the distributed HPC network

modules, it is also possible to test them. A user with authorizations has the ability to run solutions on his own or test calculation nodes, bypassing module authorization. However, these activities are significantly reduced by the appropriate limits of the resources of the computing platform to limit fraud.

A complete Distributed HPC network is composed of potentially many Computation Resources. These resources have to be organised and efficiently brokered. For this reason, we introduce additional network elements, as shown in Fig. 4. Main (Master) Nodes represent centrally managed machines which group sets of CRs and provide associated services to the end-users. The network may also contain Private Nodes which provide services only to restricted groups of users; for example for development or testing purposes. Important elements in the network are CT Brokers. They are responsible for the process of "brokerage"—assigning CTs and their CT Steps to CRs during runtime. Additional details related to this mechanism are shown in Fig. 5. It is worth mentioning that for the purpose of brokerage, CA Steps in an algorithm should be divided into CA Step Groups. These groups determine CT Steps that should be assigned to the same or closely located CRs.

4 Functional Requirements Model for Distributed HPC

Based on the DHPC vocabulary defined through the domain model we can finally define functional requirements for Distributed HPC systems. This is illustrated through a use case model presented in Figs. 6, 7 and 8 Computation Application

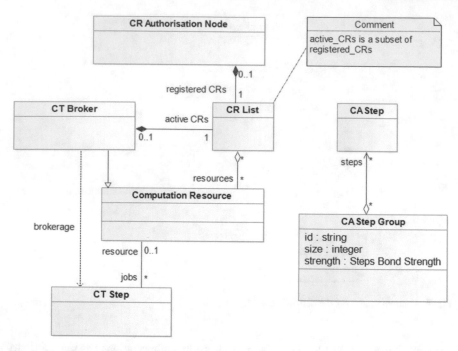

Fig. 5 Domain model for brokerage of HPC computations

development is centred around Module Shelves ("Show Module Shelf", see Fig. 6). Independent developers can develop their CMs and CAs, and test them using test CRs ("Configure test CRs" and "Run Module in test mode"). To be used within the computation network, a module needs to be placed in the central Module Repository and approved ("Submit Module" and "Approve Module"). Developers can access and use these approved modules when developing their applications ("Edit CA"). As the pool of Computation Modules grows, building a Computation Application tends towards composing flows of ready blocks of computation algorithms.

All the centrally approved CAs are accessible to end-users through the App Store ("Browse Application Store", see Fig. 7). The users can browse the store and place CAs in their local Application Shelfs ("Add Application to Application Shelf"). To run a CA, the user opens a Computation Cockpit and starts a Computation Task ("Create a CT"). It has to be noted that running a CA differs to running a typical interactive application. It is normally composed of a rather small amount of user interaction (entering parameters) followed by relatively long periods used to run specific CMs on assigned CRs. Thus, end-user interaction would become more asynchronous. Usually, the end-user would enter most of the computation parameters at the beginning ("Set CT Execution Schedule"). However, sometimes a user interaction might be needed (as determined by the application flow) somewhere in the middle—between CM calls. In such cases, the Computation Cockpit needs to inform the user about the necessity to interact and allow to perform such interaction

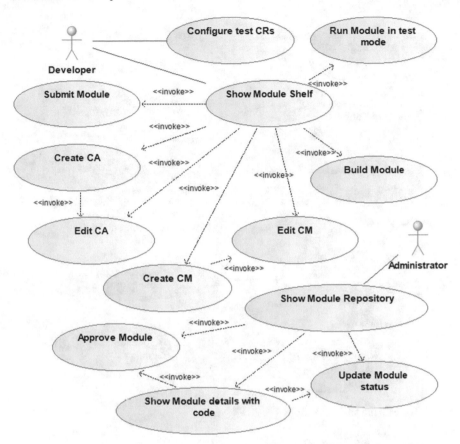

Fig. 6 Use case model for CA development

whenever the user is ready ("Interact with CT"). It can be noted that in such situations, the CAs would need to be put in a "sleep" mode in order to save computation resources.

To illustrate the functionality required from a DHCP system, as seen by the end-user, below we present a detailed scenario for the "Create a CT" use case. This use case is central from the point of view of further development of the underlying distributed computation mechanisms.

```
1. E-u selects Computation Application
2. E-u selects ''Create Computation Task'' button
3. System shows ''CT Start'' screen
4. E-u enters Computation Valuation
5. System invokes ''Set CT Execution Schedule'' use case
6. E u selects ''Confirm'' button
7. System initiates Computation Task
```

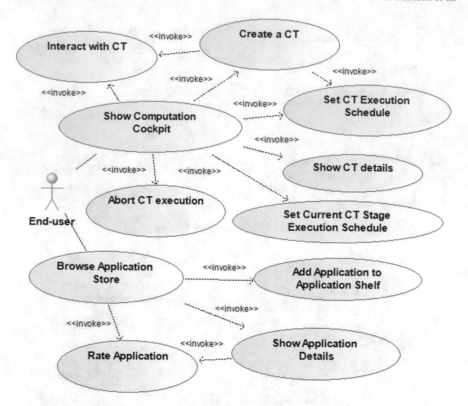

Fig. 7 Use case model for CA usage

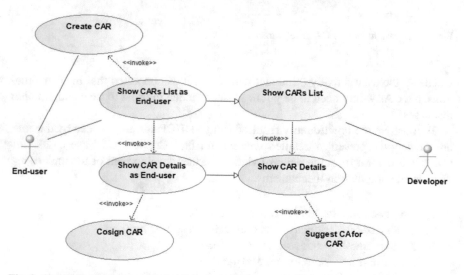

Fig. 8 Use case model for managing CA requests

```
8. [initial interaction with the user needed]
   System invokes ''Interact with CT''
9. System executes Computation Task
```

As we can see, the end-user has to define the Computation Valuation data that determines priority of computations. It can be noted that the Computation Task already contains the manifest data that can be used by the DHPC system to distribute the computations to appropriate Computation Resources. If needed, the "Set CT Execution Schedule" use case allows to enter all the Data Source Addresses and Parameters in advance. Alternatively, appropriate interactions will need to be invoked ("Interact with CT") when required by the application code. After possible initial interaction, the system executes the Computation Task. Inside the system, this means that individual Computation Steps are automatically assigned to matching Computation Resources. This is done in accordance with the sequence and parallelisation of these steps within the algorithm of the particular CA.

Finally, the end- users also have the possibility to specify details of applications that are currently unavailable, but could be useful to them ("Create CAR", see Fig. 8). Such CARs (Computation Application Requests) are also browsable by other users ("Show CARs List as End-User", "Show CAR Details as End-User"), which have the possibility to co-sign them ("Cosign CAR"). Finally, when popular requests are implemented by the developers, they have the possibility to suggest their application as a solution to a specific CAR ("Sugest CA for CAR").

Another set of requirements associated with management and supervision of CRs is shown in Figs. 9 and 10. The end-users can access their CRs trough browsing owned CR Groups ("Show owned CR Group list", see Fig. 9). Then they have the possibility to execute certain actions an CR groups ("Submit a CR Group", "Safely deactivate a CR Group") or go to managing a specific CR ("Show own CR Group Resource list").

Supervision of CRs is based around a list containing all CR Groups ("Show all CR Group list', see Fig. 10"). Then, the Administrator can change resource statuses ("Activate/deactivate a CR Group") or check the details of particular CRs ("Show CR Group Resources list", "Show Advanced CR details") and run benchmarks on them ("Execute Single Benchmark").

5 Reliability and Security of Calculations

As a supplement to the functional requirements, we can define key quality characteristics required from a Distributed HPC system: reliability and security. To assure high degree of reliability, practically every element of the computing platform needs to be duplicated or replaced by a spare element. For this purpose, additional units have to be established for all network elements. In addition, as part of the computing platform, a mechanism needs to be designed to allow automatic reconstruction of

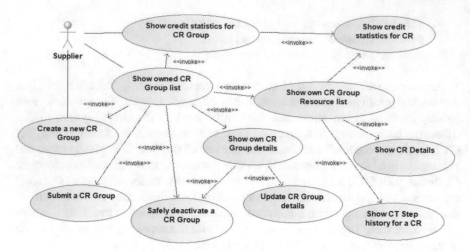

Fig. 9 Use case model for CR management

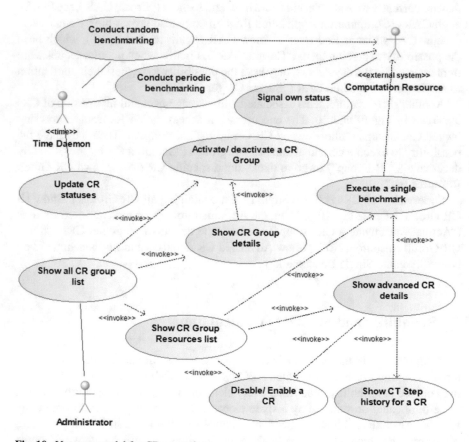

Fig. 10 Use case model for CR supervison

connections between network nodes in such a way as to maintain the consistency of the computing network.

The system, in its assumptions, needs to implement the transparency principle in the form of:

- **transparency of access**—unification of data access methods and hiding differences in their representation between computation modules,
- **position transparency**—users cannot determine the physical location of the resource except for the input data and the location of the calculation results,
- **transparency of migration**—the ability to transfer resources between servers without changing the ways of referring to them in the event of a computing node failure or termination of its availability,
- **transparency of movement**—resources can be transferred even during their use, primarily when calculations are performed in parallel,
- **transparency of reproduction**—hiding users from multiplying resources in the form of duplicated nodes,
- **transparency of concurrency**—the possibility of concurrent data processing does not cause inconsistencies,
- **failure transparency**—masking transient failures of individual components of the distributed system by transferring calculations to another calculation node,
- **transparency of durability**—masking the method of storing the resources in the computation network.

To achieve the desirable level of security, a unique mechanism for the authentication of computing modules needs to be designed. Each module created by the programmer must be approved by placing the applications in a batch of applications by the network administrators. The credibility of the calculation modules will be confirmed by the appropriately implemented hierarchical structure of the electronic signature. In addition to the security of the software, an appropriate procedure is proposed for adding new nodes to the computing network. The node must be manually added by network administrators, while its manifest must be prepared by the local node administrator (participant of the network) and verified by appropriate tests run on the newly added node. The node's data will also be confirmed by an appropriate electronic signature.

In addition to the security of calculation modules, it is necessary to ensure data security in the computing network. In order to protect data appropriately generated by computing modules, a blockchain-based blocking mechanism is envisioned. In this way, it will be possible to ensure reliability and integrity of data flow within the entire computing network. The blockchain mechanism will also be used to ensure settlements between participants of the computation network.

6 Summary and Future Work

Analysis of existing literature on DHPC domain models and functional requirements shows lack of direct approaches of this kind. Our main contribution is thus the presented requirements model, which is aimed at enabling the development of more coherent and universal DHPC systems in the future. We provide a unified vocabulary and functional characteristics, including a use case model with example scenario details that can be applied to a wide range of future DHPC systems. The presented models thus can serve as a unifying framework to join a heterogeneous set of Computation Resources into a coherent computation environment. Such a system would involve a set of centrally managed machines (Master Nodes) that would allow to manage the development of Computation Applications, run Computation Tasks and manage Computation Resources. As future work, we also plan to develop a model for handling payments within the system. We should note that all the information required for this purpose is already available in the presented model. The challenge here will be to develop an appropriate scheme and technology that would take care of such distributed payments.

The core network of Master Nodes would be complemented by a growing set of Computation Resources, offered by various suppliers. Further research agenda would thus need to involve the development of a unified architecture (both physical and logical) for this new system. Challenges in this area involve designing a runtime engine for CAs that would involve brokerage of CMs as CT Steps.

The runtime engine would need to handle the execution of CA Steps, including CM calls and user interactions. This would need to be based on building a high-level language to define sequences of CA steps—we call it the Computation Application Language (CAL). We can observe that the presented domain model can be quite naturally transformed into a metamodel for CAL. This would allow to treat it as a domain-specific language and create a model-driven framework—a visual editor and code generator. The challenge here is to assure ease of use (even for non-IT personnel) and necessary levels of computation performance of the runtime engine for CAL. The Computational Application Language proposed by us can be adapted for any field, which allows the implementation of a domain-based, high-performance distributed computing platform.

Building a full computing environment based on the proposed solution would allow for performing highly-scale calculations in a fully distributed computational model using multiple computing nodes. Such a solution can be implemented both for large computing centres and for small teams that need short-term access to high computing power. The proposed solution may be in the future a base for creating a commercial computing network using distributed computing nodes. The designed solutions allow to scale the solution to various types of problems efficiently, and making them independent of the fixed domain allows for simple adaptation to any applications.

The above-introduced research agenda is currently underway as part of the BalticLSC project where the goal is to develop and validate the system consistent with the presented conceptual requirements models.

Acknowledgements This work is partially funded from the European Regional Development Fund, Interreg Baltic Sea Region programme, project BalticLSC #R075.

References

1. Armstrong R, Gannon D, et al (1999) Toward a common component architecture for high-performance scientific computing. In: Proceedings. The eighth international symposium on high performance distributed computing. IEEE Computer Society, pp 115–124
2. Bernholdt DE, Allan BA et al (2006) A component architecture for high-performance scientific computing. Int J High Perform Comput Appl 20(2):163–202. https://doi.org/10.1177/1094342006064488
3. Blaze M, Feigenbaum J, Ioannidis J, Keromytis AD (1999) The role of trust management in distributed systems security. In: Secure internet programming. Springer Berlin Heidelberg, pp 185–210
4. Blaze M, Feigenbaum J, Lacy J (1996) Decentralized trust management. In: Proceedings 1996 IEEE symposium on security and privacy. IEEE, pp 164–173
5. Bryant BR, Gray J, Mernik M (2010) Domain-specific software engineering. In: Proceedings of the FSE/SDP workshop on future of software engineering research. ACM, pp 65–68. https://doi.org/10.1145/1882362.1882376
6. Bunch C, Chohan N, et al (2011) Neptune: a domain specific language for deploying HPC software on cloud platforms. In: Proceedings of the 2nd international workshop on scientific cloud computing - scienceCloud 11. ACM Press, pp. 59–68. https://doi.org/10.1145/1996109.1996120
7. Cappello F, Djilali S, Fedak G, Herault T, Magniette F, Néri V, Lodygensky O (2005) Computing on large-scale distributed systems: xtremWeb architecture, programming models, security, tests and convergence with grid. Futur Gener Comput Syst 21(3):417–437. https://doi.org/10.1016/j.future.2004.04.011
8. Dongarra J, Sterling T, Simon H, Strohmaier E (2005) High-performance computing: clusters, constellations, MPPs, and future directions. Comput Sci Eng 7(2):51–59. https://doi.org/10.1109/mcse.2005.34
9. Foster I, Zhao Y, Raicu I, Lu S (2008) Cloud computing and grid computing 360-degree compared. In: 2008 grid computing environments workshop. IEEE, pp 1–10. https://doi.org/10.1109/gce.2008.4738445
10. Giles MB, Reguly I (2014) Trends in high-performance computing for engineering calculations. Philos Trans R Soc A: Math, Phys Eng Sci 372. https://doi.org/10.1098/rsta.2013.0319
11. Grimshaw AS, Wulf WA et al (1997) The legion vision of a worldwide virtual computer. Commun ACM 40(1):39–45. https://doi.org/10.1145/242857.242867
12. Hager G, Wellein G (2010) Introduction to high performance computing for scientists and engineers. CRC Press
13. Hernández F, Bangalore P, Reilly K (2005) Automating the development of scientific applications using domain-specific modeling. In: Proceedings of the second international workshop on Software engineering for high performance computing system applications - SE-HPCS 05. ACM Press, pp 50–54. https://doi.org/10.1145/1145319.1145334
14. Lampson B, Rivest R (1997) Cryptography and information security group research project: a simple distributed security infrastructure. Technical report, Technical report, MIT

15. Laure E (2000) Distributed high performance computing with OpusJava. In: Parallel computing: fundamentals and applications. Published by Imperial College Press and Distributed by World Scientific Publishing Co., pp 590–597. https://doi.org/10.1142/9781848160170_0070
16. Laure E (2001) High level support for distributed high performance computing. Ph.D. thesis, University of Vienna
17. Li Y (2015) DRUMS: domain-specific requirements modelingfor scientists. Ph.D. thesis, Technische Universität München
18. Li Y, Guzman E, Bruegge B (2015) Effective requirements engineering for CSE projects: a lightweight tool. In: 18th international conference on computational science and engineering. IEEE, pp. 253–261. https://doi.org/10.1109/cse.2015.49
19. Liu H, Parashar M (2005) Enabling self-management of component-based high-performance scientific applications. In: 14th IEEE international symposium on high performance distributed computing, HPDC-14. IEEE, pp 59–68
20. Membarth R, Hannig F, et al (2012) Towards domain-specific computing for stencil codes in HPC. In: 2012 SC companion: high performance computing, networking storage and analysis. IEEE, pp 1133–1138. https://doi.org/10.1109/sc.companion.2012.136
21. Merkel D (2014) Docker: lightweight linux containers for consistent development and deployment. Linux J 2014(239). http://dl.acm.org/citation.cfm?id=2600239.2600241
22. Moreira JE, Midkiff SP et al (2000) Java programming for high-performance numerical computing. IBM Syst J 39(1):21–56. https://doi.org/10.1147/sj.391.0021
23. Palyart M, Lugato D, Ober I, Bruel JM (2011) MDE4HPC: an approach for using model-driven engineering in high-performance computing. In: SDL 2011: integrating system and software modeling, vol 7083, pp. 247–261. https://doi.org/10.1007/978-3-642-25264-8_19
24. Palyart M, Ober I (2012) Other: HPCML: a modeling language dedicated to high-performance scientific computing. In: Proceedings of the 1st international workshop on model-Driven engineering for High performance and CLoud computing - MDHPCL12. https://doi.org/10.1145/2446224.2446230
25. Ranjan R, Benatallah B et al (2015) Cloud resource orchestration programming: overview, issues, and directions. IEEE Internet Comput 19(5):46–56. https://doi.org/10.1109/mic.2015.20
26. Schmidberger M, Brugge B (2012) Need of software engineering methods for high performance computing applications. In: 11th international symposium on parallel and distributed computing. IEEE, pp 40–46. https://doi.org/10.1109/ispdc.2012.14
27. Schmitt C, Kuckuk S, et al (2014) ExaSlang: a domain-specific language for highly scalable multigrid solvers. In: 4th International workshop on domain-specific languages and high-level frameworks for high performance computing. IEEE, pp 42–51. https://doi.org/10.1109/wolfhpc.2014.11
28. Teliba H, Cisterninoa M, Ruggierob V, Bernardc F (2016) RAPHI: rarefied flow simulations on xeon phi architecture. Technical report, SHAPE Project
29. Van De Vanter ML, Post DE, Zosel ME (2005) HPC needs a tool strategy. In: Proceedings of the second international workshop on software engineering for high performance computing system applications - SE-HPCS 05. ACM Press, pp 55–59. https://doi.org/10.1145/1145319.1145335
30. Vecchiola C, Pandey S, Buyya R (2009) High-performance cloud computing: a view of scientific applications. In: 2009 10th international symposium on pervasive systems, algorithms, and networks. IEEE, pp. 4–16. https://doi.org/10.1109/i-span.2009.150

Requirement Engineering as a Software Development Process

Pawel Baszuro and Jakub Swacha

Abstract The paper introduces a novel requirements management tool, incorporating software development practices into requirements engineering. It is motivated by an expectation that using well-established techniques should provide benefits such as increase of requirements quality, better alignment of tooling to organization needs and support for Agile project management techniques. Our aim, therefore, is to provide an open architecture for various requirements engineering activities, as till now, many of existing requirements management tools do not follow the latest developments in software development practices. In this paper, we propose a mapping of various aspects of software development based on Continuous Integration to requirements engineering and describe a prototype requirements management tool built to validate the proposed concept.

1 Introduction

There are several criteria established for software requirements, that include completeness, consistency, traceability, and testability [1]. More than two decades ago, Standish Group published results of a survey where factors related to requirements (number 1. Incomplete Requirements and number 6. Changing requirements and specifications) were in top 10 on "project impaired list" [2], and became important factors on shaping requirements engineering research [3].

Requirements engineering in modern project methodologies can be considered as challenging. Agile Manifesto values "working software over comprehensive documentation" [4], a sentence that is translated into "not enough documentation" that can

P. Baszuro (✉)
Bugolka, Warsaw, Poland
e-mail: pbaszuro@acm.org

J. Swacha
Institute of Information Technology in Management, University of Szczecin, Mickiewicza 64,
71-101 Szczecin, Poland
e-mail: jakubs@uoo.univ.szczecin.pl

© Springer Nature Switzerland AG 2020
A. Poniszewska-Marańda et al. (eds.), *Data-Centric Business and Applications*,
Lecture Notes on Data Engineering and Communications Technologies 40,
https://doi.org/10.1007/978-3-030-34706-2_2

be used for future reference [5]. Close cooperation between various groups of stake-holders and the project team is a key benefit in Agile methodologies. Knowledge sharing for future reference must be considered a lightweight process to produce "minimal specification" that can be input to prototype or solution [6]. In order to satisfy this lightweight process, different forms of requirements artifacts were intro-duced to establish high-level goals and expectations on software. Selected ones are User Stories and Acceptance Criteria. User Story is a short phrase written in natural language that links an actor, feature, and benefit. Typically User Story is expressed as the sentence "As an [Actor] I want to [feature] so that [benefit]" [7]. Such a phrase may be ambiguous to other stakeholders or project team, so often User Story is paired with one or more artifacts in the Acceptance Criteria format. Acceptance Criteria can be expressed in natural language, but structured to three steps: Given, When, Then. "Given" sets up the initial context for the requirement, "When" describes the event that occurs, and "Then" elaborates on system expectations. Requirements expressed in Acceptance Criteria format can be covered by automated tests using Behavior Driven Development (BDD) tools [8]. BDD is one of the material touch points between requirements engineering and software development. Requirements artifacts in the form of Given-When-Then sentences are copied into test content and committed to the source code base.

In this paper, we define key design goals and a model of requirements management platform aimed at Agile practitioners to bridge the gap between source code tooling and requirements engineering, which includes:

- definition of key terms on the requirements engineering platform,
- organization of proposed platform,
- the concept of requirements quality measures, and
- technology of implementation.

A new requirements engineering platform conforming to this model is being developed to serve as a blueprint for future platforms.

2 Related Work

There are many attempts to increase the quality of software requirements, from both requirements engineering and tooling perspectives. In the quest to answer needs such as the ease of use, facilitating negotiation process, synchronous collaboration, and versioning, wiki-based solutions are being researched. These solutions address many challenges such as the involvement of stakeholders, collaboration and inter-disciplinary teams [9]. In project teams, multiple different groups may be involved in shaping requirements, not only requirements engineers but first of all clients and users. There are vast differences among them, which include [10]:

- different perspectives on the system under development;
- different backgrounds, which can cause communication problems;

– different objectives, which influence views on the requirements;
– different abilities to express requirements and document them using a technical platform;
– different involvements.

The open nature of editable pages supports incremental evolution and collaboration [11] as well as negotiation of requirements [9]. The wiki-based solutions also support interdisciplinary project teams [12]. The challenge of low quality requirements must be addressed by skilled users, who have to reorganize existing imperfectly defined requirements, though a wiki system can support this process by automatically producing feedback to stakeholders who provide poor specifications [11].

In order to tackle the problems with textual descriptions, the use of linguistics patterns for concept description was proposed [13]. There are also attempts for documenting requirements in more strictly structured forms, such as YAML [14] or SysML [15].

Different requirements management tools are analyzed for features such as requirements analysis, links and traceability, change management, lightweightness, tool integration, document generation and price [16]. It is worth to mention that none of the surveyed tools (both commercial and open source) provides all the specified features at the same time.

In order to address specific aspects of requirements engineering, specialized tools are proposed. For the process of requirements traceability checking, new languages are proposed, including: TRL [17], TQL [18], and TracQL [19]. Traceability is a key concern for tracking lifecycle of the requirement within various types of artifacts. TRL stands for A Traceability Representation Language. It provides an abstraction to requirements and links, same as queries running on top of the requirements [17]. TRL tracks links as a central artifact in the whole traceability process. TraceLink has its name, same as kind of relation (e.g. dependency, refinement, etc.), link to requirement and actual artifact of given type. To successfully evaluate queries, Query Abstract Syntax is proposed. In comparison to TracQL, during evaluation of a query trace links using TRL, there is no need to traverse entire graph of dependencies. TQL is XML like language, where requirements are nodes, and links are locators of nodes. TQL queries work using XPath query language over XML namespace. All artifacts (source code files, requirements, etc.) are converted to the XML format to run TQL queries on top.

Requirements engineering tools may have text analysis embedded that performs quality assurance of requirements (e.g. detection of subjective language, ambiguous wording, loopholes, incomplete references) [20]. There are attempts to automate verification process using SysML [21, 22] or model-based techniques [23, 24]. Requirements validation is addressed in multiple ways: by expressing requirements in terms of BNF notation, then parsing it and test case modeling [25], or by direct natural language processing [26].

3   Key Terms of Requirements Engineering Platform

The proposed platform supports various activities in the requirements engineering process. We start from defining key terms and then describing relevant software development activities, that are mapped against each other.

3.1   Key Terms: Requirements Engineering Perspective

For the purpose of this paper, we will consistently use the definition of terms from Business Analysis Book of Knowledge (BABOK) [27]. The proposed platform is to meet conditions established for requirements management tool with the capabilities listed in Table 1.

3.2   Key Terms: Software Engineering Perspective

Modern software engineering benefits from Continuous Integration practice that establishes the continuity between software development and its operational environment [29]. Relevant terms are continuous integration, continuous delivery, continuous verification, and continuous testing. Relevant terms are defined in Table 2.

4   Requirements Engineering Platform Perspective

The proposed platform is based on a novel approach, combining several requirements engineering and software engineering practices. The practices are divided into artifacts-related and process-related ones. Elaboration of both is put below.

4.1   Project Organization on the Platform

Similarly to the source code, related requirements are organized in a requirement project. Below, the project file types are described and it is explained how usage version control system and simple text transformations can be used to support requirements engineering activities.

Each requirement project includes three sets of documents: templates, artifacts, and metrics definitions. The proposed platform combines multiple approaches to address traceability problems. It uses field identification (via conversion rule that is

Table 1 Capabilities of the requirements management tool, based on [27]

Capability	Definition
Elicitation and collaboration	The process of gathering information from stakeholders and the process of two or more people working together. People coordinate their actions to achieve a goal [28]
Requirements modeling and specification	Activities that define some of the aspects of the state (current or future)
Requirements traceability	The function of tracking relationships between artifacts
Versioning and baselining	Versioning is a process of assigning a version to a particular requirement and its change. A process of requirements reviewing and changes approving
Attribute definition for tracking and monitoring	A process of setting up attributes for tracking and monitoring purposes (i.e. capturing requirements status or relationship within different aspects)
Document generation	Ability to convert particular requirement artifact (or multiple of them) to another format (i.e. word processor document or web format)
Requirements change control	A set of activities that rely on requirements versioning and baselining, ensures that any changes to requirements are understood and agreed before they are made. Requirements change control may include requirements verification and validation activities
Requirements verification	An activity that ensures that requirements are defined correctly with an acceptable level of quality
Requirements validation	An activity that checks if requirement supports the expected goal

similar to XPath), temporary database creation and evaluating various queries to identify requirement quality issues. The template is a technical description of a requirement artifact. The template has the following attributes: unique identifier, humanfriendly name, description, optional base template, technical conversion template, set of verification rules and set of validation rules. The template can be described using elements of source code (e.g. regular expressions, paths, queries). The artifact is an actual requirement. Each requirement has its type, unique identifier, and content. Type is a reference to the relevant template. Content must have a structure as defined in the referenced template. Each artifact must be a text document (in relation to the source code for popular programming languages), in the format that is easy to parse and process. Artifacts may contain links to other artifacts. Metrics are macro-control rules that are to be run against all artifacts in the project (in comparison to template validation rules that focus on a particular artifact). Each document describing either

Table 2 Software engineering continuous practices

Continuous practice	Definition
Continuous integration	A process that is triggered automatically and consists activities of obtaining the latest version of source code, compiling the code, running unit and acceptance tests, validating code coverage, checking code standards and building deployment packages (the process is visualized on Fig. 1). Automation is required to run the process on a regular basis to give quick feedback to developers about potential issues
Continuous delivery	A practice of preparation package of software that successfully passed the Continuous integration process, that can be accessed by users (not always deployed to actual users)
Continuous verification	A set of formal methods and inspections for early issue detections rather than relying on testing
Continuous Testing	A process of automated testing, that helps to reduce the time between the introduction of errors and their detection

a template, an artifact, metrics or any asset gathered during requirement elicitation must be a file.

As all relevant documents form separate files, it is possible to store them separately in a repository managed with a version-control system. Version-control systems offer the functionality of tracking changes on individual files. Each change has attributes of its time, author, and reference to its previous version. Having version control systems allow recovering previous versions of the file whenever mistake occurs. Moreover, many version control systems have the function of branches that group related changes into baskets to allow separation of work being done in multiple workstreams. Versions (even from multiple branches) can be joined with the merge function. Prior to merging, the latest version of both files is fetched, checked for differences, merged and potentially merge issues resolved and then submitted to target branch. There could be more advanced processes around merging, including skilled person review. Modern version control systems (like e.g. Git) offer distribution of file structure and history to mirror repositories on workstations that help collaboration between project team members.

Having requirements artifacts expressed in text files allows end-users to use any tool to elicit and model requirements. Depending on the users' skills, in order to edit an artifact, each user may use simple system tools like vi or emacs, or web-based interface to the version control system. The use of the version control system provides a platform that satisfies conditions of versioning. Branches can be used as a way to segregate requirements. Segregation can be done on multiple axes, business-perspective- or process-perspective-related. Business perspective can be related to the offering of the same component with massive customizations for a particular application. Process perspective allows segregation of work in progress and work reviewed and accepted. Process-perspective segregation of branches is in relation to requirements baselining (i.e. merge from work in progress branch to the reviewed one).

4.2   Requirements Engineering Processes on the Platform

This section describes the Continuous Integration process applied to requirements engineering on the platform, where source code compiling is replaced by requirements processing. Description of the requirements processing and mapping to the requirements engineering capabilities is elaborated below.

As requirements are stored in a version control repository (VCS), then on the proposed platform they are subject to continuous integration. Requirements are fetched from a VCS and processed by the platform. Requirements processing has the following steps: processing project file, processing all templates in the project, processing all artifacts in the project, and then calculating metrics.

Processing project file step is intended for reading project file content and identifying all relevant files included in the project file (i.e. there could be files in VCS that are not subject for processing, i.e. graphics files). Processing all templates in the project is a step in which all the requirements template files referenced in the project file are read and parsed. The template may have additional attributes like a base template or abstract flag. Base template attribute contains an identifier of another template that provides base structure to the given template. It is the role of the platform to resolve the chain of dependencies and combine all defined sections (technical conversion template, set of verification rules and set of validation rules) into one artifact during processing templates runtime. The abstract flag means that particular template cannot be referenced by an artifact, but can be used to provide common sections to derived templates.

When all templates are successfully resolved, then requirement artifacts processing occurs. Same as templates, all relevant requirement artifacts are read from files in the project and processed one by one. Processing a single requirement is a process having the following steps: reading the file, parsing its structure, choosing a template, all of them based on template processing with template rules applied in three steps. First, the structure is mapped using a technical conversion template to the analysis structure. The outcome, the analysis structure provides a set of relevant attributes for the subsequent steps. Second, the analysis structure is verified using verification rules from the template. Verification checks if all attributes from analysis structure meet given criteria: is not empty, its value meet additional conditions for constant values (like mandatory word for "Given" for pre-amble in Acceptance Criteria defined in Given-When-Then form), string values (i.e. content must have at least three characters), number variables (i.e. values are non-negative) or special values (like Universal Resource Identifiers, URI). The outcome of step two is a verified artifact and relevant relationships stored in the traceability database. A simplified diagram of traceability database is presented on Fig. 2. As can be seen, each artifact is associated with artifact template. Artifact template may have base template used for inheritance (i.e. one artifact template may extend another artifact template). Links are extracted connections between artifacts. Each link has source and destination artifact.

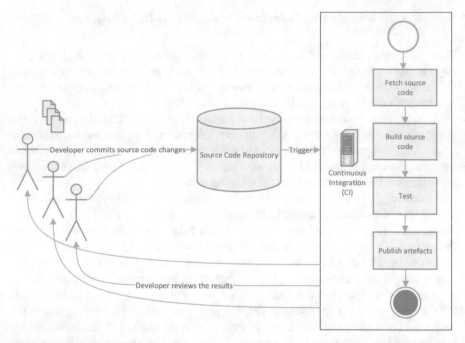

Fig. 1 Visualization of continuous integration with relation to the source code repository and continuous processes running as part of continuous integration

Third, the artifact is validated using validation rules. Validation rules run on top of verified document and relationships from traceability database. Validation rules are to check expectations on a particular requirement (i.e. a User story starting with "As a <user>" in the field <user > must contain a link to an artifact with the type of Actor). Each validation rule is a mathematical logic predicate. Each predicate produces either true (success) or false (failure) as a result of the evaluation of validation rule. When failure is produced, then relevant error message associated with validation rule is printed to the output. Failures associated with any template or artifact processing are combined to the single output. If none are present, then it means that processing succeeded. Metrics are calculated based on data gathered in traceability database and printed to the output. Successfully processed artifacts optionally can be converted to the desired form (i.e. a website with interactive links) and published to an external source.

The overall structure of Requirements Template in terms of syntax is presented on Fig. 3. It shows the template with three types of rules: conversion rule, verification rule and validation rule. Validation rule is associated with a validation rule query (query is run on the traceability database, see Fig. 2). Validation Rule query contains collection of tables (identified by unique alias) and query parameters. Parameters are ordered in Validation Rule Query Clause (translated to the "where" clause in

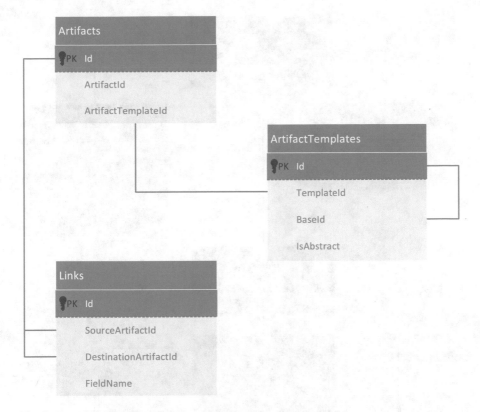

Fig. 2 Traceability database diagram with artifacts, templates and links

SQL), alias must be reflected in artifact table, and field is field name (see structure of traceability database on Fig. 2).

In this section, we mapped various aspects of the proposed platform that support various conditions for the requirements management tool. They are as follows:

- requirement modeling and custom attribute tracking processes tailored to the needs of the particular organization with the flexibility of changes at a point in time (modification of requirements template);
- requirements traceability that can be evaluated from the artifact and metrics perspectives;
- requirements verification and validation as key elements of ensuring the quality of single artifacts;
- the ability to evaluate any change to the project, templates or requirements ensuring that any issues would be detected early (in line with continuous verification and continuous testing approaches).

The entire process of automated requirements processing is in line with continuous development approach from software engineering.

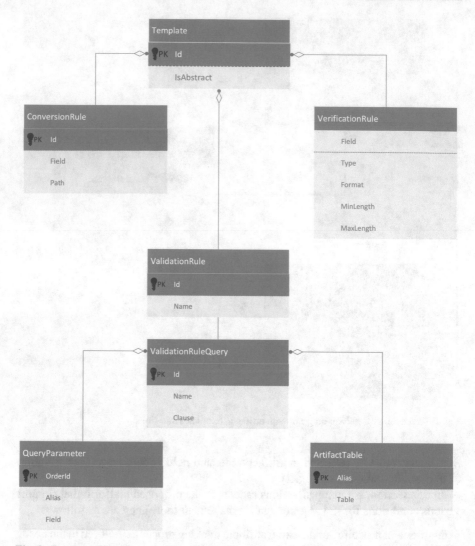

Fig. 3 Structure of requirements template in terms of syntax

5 Requirement Quality Measures

Quality measures are aligned across verification and validation artifact rules, and metrics running on top of all artifacts collected.

5.1 Verification Rules

The role of verification rules is to check the structure and format of requirement. Each artifact template may contain one or more verification rules, each one associated with a field. Verification rule is a rule that runs on top of an analysis structure. Analysis structure is created from the parsed file (a tree-like structure) and mapped using traversal to relevant fields.

The definition of a verification rule associated with a field is to provide the type of expected data and its constraints (i.e. string format or minimal length) and check field value against it.

5.2 Validation Rules

Validation rule supports traceability goal on the platform by checking the relationship between given artifact and the rest of the artifacts within the project. The role of a validation rule is to provide input for construction of predicate function. Predicate function operates on verified analysis structure and content of traceability database. Predicate function runs a parameterized query to answer the simple question (i.e. if particular requirement traces correctly another requirement).

5.3 Metrics

Macro metrics provide a big picture of the requirements project. Metrics form a set of rules running on top of traceability database. Each metric rule is composed as a non-empty set of sub-queries. It may return a single value (i.e. the number of requirement artifacts in the type of Actor) or a table (e.g. the top ten of the most linked artifacts).

6 Application and Technology of Implementation

This section covers the application of the proposed approach and its technology of implementation. A prototype of the proposed platform can be used in two scenarios. First, the user may use the platform to draft new requirements prior to sharing it with the rest of the project team. Evaluation of requirements is done as in any standalone application, locally to the user and in isolation of changes with changes from the other user. Secondly, the user may commit changes to the files, share it, and run process run verification and validation processes. Figure 4 presents this case, whereas Fig. 5 contains visualization of the requirements process within the pro-

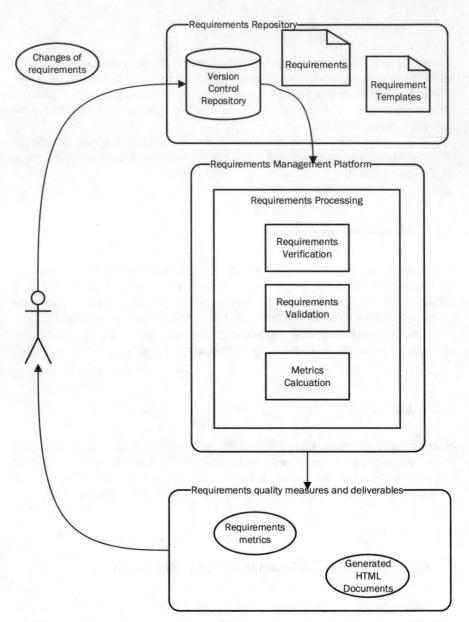

Fig. 4 Proposed platform architecture diagram with requirements flow, repositories, processes and objects used

posed requirements management platform, exposing details of processing requirements on the platform that contains steps of parsing requirement file (User Input), parsing requirement template, processing structures, requirement template structures (derived from requirement template) and generated document with validation rules output.

User may commit the change they made to either dedicated feature or component branch or to the main branch in the repository. Working on component or

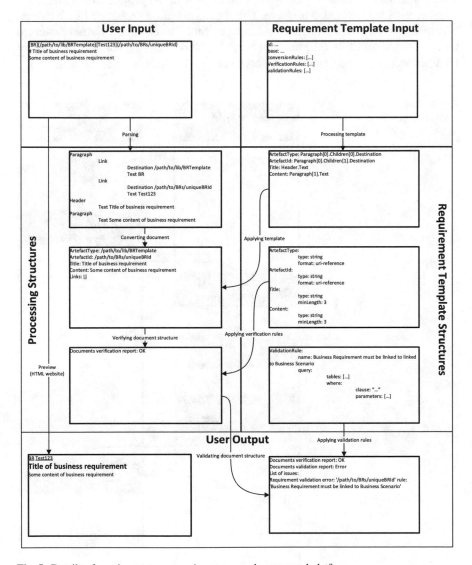

Fig. 5 Details of requirements processing steps on the proposed platform

feature branches allows users to work independently, where all requirement metrics are generated only on base, changed or newly created requirements. Separation of requirements metrics allows detecting any functions that are not fully covered by requirements. In connection to other deliverables in the repository, such as source code or test scripts, a comprehensive view of entire solution is provided end-to-end, from requirements through actual implementation to test scripts. Merging changes from a branch to the main branch, or working directly on the main branch, gives the ability to align impacted requirements with the rest of the project. During merge, a requirements review is also possible. The activity diagram shown on Fig. 6 depicts control flow with various activities within repository. The order of activities presented therein comprises to start the feature branch by "fork" activity, multiple commits of changes on the feature branch, and merge back to the main (master) branch.

Fig. 6 Activities on requirements within repository

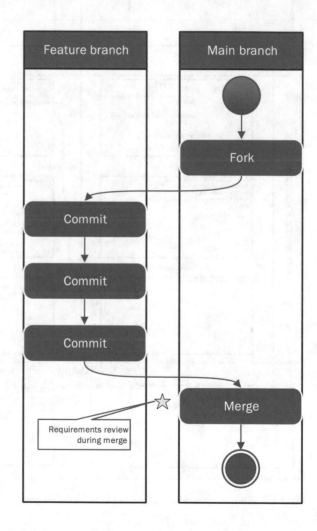

```
/**
 * Sets the value of variable to given number
 * @param {Number} number must be an integer
 *
 * Requirements at {@link
https://github.com/reqslang/reqslang/blob/4ec0aee1c...}
 */
setTo(number) {
    this.variable = number;
}
```

Fig. 7 Snippet of source code in TypeScript featuring a link to requirements

Each commit has a unique identifier (i.e. version number or hash value) that can be used to identify the state of the requirement at any given point in time. Given the path to the requirement with inclusion of its version may be used in source code to associate it to the related requirements. Many documentation generators for modern programming languages provide sets of dedicated tags that could be used to embed metadata directly in the code base. For example, in Java programming language there is a Javadoc generator, that allows automatic generation of documentation based on comments in the source code. Javadoc has a tag called "see", that is used to provide links to other documents (another entity in code base or hyperlink to external document via HTML anchor element). The external links contained in the comments within the code base may point to the uniquely identified requirements. Figure 7 presents a snippet of source code written in TypeScript, a fragment of a class containing function called "setTo". The "setTo" function has documentation in code base for documentation generator. The "link" tag contains a URL pointing to requirements committed with the hash identifier given there (starts with value "4ec0aee"), followed by the file path (which is truncated on the figure due to limited page width).

After the commit of a requirement, there is a way to spot the impacted portion of the codebase, thus to have an impact analysis initiated automatically. The results of the impact analysis can then be put under further review by a skilled person. Moreover, as an additional step in the post-build stage of the requirements processing, the text comparison of relevant artifacts can be performed. For acceptance criteria, with the requirements artifacts specified in Given-When-Then format, it is possible to export requirements as plain text (i.e. without any mark-up) and compare it to the relevant parts of the tests specified in BDD format (i.e. the same Given-When-Then parts of the test feature files). Similarly documented use cases, with a chain of user actions denoted in requirements specification, can be cross-checked with scenarios defined in model-based testing test scripts. As most testing tools offer the capability of having comments contained in the test scripts, by using the same pattern in comments section consistently, the test script elements can be linked to respective requirements. The

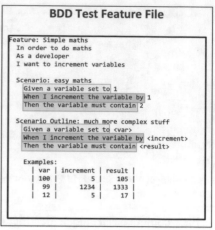

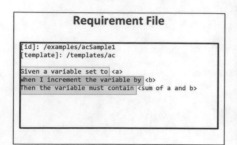

Fig. 8 Exemplary mapping between requirements file on the proposed requirements management platform and test feature file

additional checks could be made in order to automatically detect requirements not covered by any tests, or tests related to decommissioned requirements.

Figure 8 shows an exemplary mapping between requirements and tests: on both the presented files, the relevant parts are highlighted. In the presented example, one requirement is covered by two test scenarios, that result in four test cases executed (one defined explicitly, the remaining ones defined with an examples table having three rows).

In order to implement a prototype of the proposed platform, a JavaScript-based stack is used. The prototype software runs in Node.JS run-time environment featuring multiple open source libraries. Project files, template files, and metric rule files are stored in the JavaScript Object Notation (JSON) format [30]. Requirements artifacts are stored in CommonMark file format (lightweight markup language) and parsed using the CommonMark library [31]. Template mapping is done using JSON-Path library (similar to XPath for XML files) [32]. The analysis structure is JSON structure that is subject to schema validation using the ajv library (a process similar to validation using of XML Schema) [33] during the requirements verification step. Traceability database is a sqlite3 in-memory database [34]. Both requirements validation rules and metric rules are evaluated using the knex library which generates SQL queries based on given query structure [35]. HTML documents are generated using the CommonMark library [31].

7 Conclusion and Future Work

The proposed approach takes software engineering practices and applies them to requirements engineering. The sketched platform satisfies multiple conditions set up for modern requirements management tool using techniques borrowed from software engineering continuous integration process.

The goal is to have a lightweight tool that can be run as part of continuous integration in the same way as desktop software bundle (user can check changes before committing it to the repository). Wiki-like format of requirements may be beneficial to beginners in requirements engineering. It also makes it feasible to analyze data gathered from various sources of project information (requirements repository, source code repository, and issue tracking system) to identify collaboration patterns within a project team [36]. On the downside, requirements template modeling based on the proposed platform requires knowledge of programming concepts (such as object-oriented modeling, structured queries, and JSONPath).

A prototype of the proposed platform is soon to be published as an open source tool available for the public audience for evaluation in a real-life working environment [37].

References

1. Boehm B (2000) Requirements that Handle IKIWISI, COTS, and rapid change. Computer 99–102. https://www.doi.org/10.1109/2.869384
2. Standish Group (1994) CHAOS Report
3. Sillitti A, Succi G (2005) Requirements engineering for Agile methods. Engineering and managing software requirements. Springer, Berlin, Germany, pp 309–326. https://www.doi.org/10.1007/3-540-28244-0_14
4. Beck K, Beedle M, van Bennekum A, Cockburn A, Cunningham W, Fowler M, Grenning J, Highsmith J, Hunt A, Jeffries R, Kern J, Marick B, Martin R, Mellor S, Schwaber K, Sutherland J, Thomas D (2001) Manifesto for agile software development. https://agilemanifesto.org/
5. Paetsch F, Eberlein A, Maurer F (2003) Requirements engineering and agile software development. In: Proceedings twelfth IEEE international workshops on enabling technologies: infrastructure for collaborative enterprises, WET ICE 2003, pp 308–313, Linz, Austria. IEEE. https://doi.org/10.1109/ENABL.2003.1231428
6. De Lucia A, Qusef A (2010) Requirements engineering in agile software development. J Emerg Technol Web Intell 3(3):212–220. https://doi.org/10.4304/jetwi.2.3.212-220
7. Cohn M (2004) User stories applied: for agile software development. Addison-Wesley, Boston, MA, USA
8. North D (2006) Introducing BDD. Better Software 3
9. Yang D, Wu D, Koolmanojwong S, Brown W, Boehm B (2008) WikiWinWin: a wiki based system for collaborative requirements negotiation. In: Proceedings of the 41st Annual Hawaii International Conference on System Sciences (HICSS 2008), pp. 24–24, ACM, Waikoloa, HI, USA. https://doi.org/10.1109/HICSS.2008.502
10. Decker B, Ras E, Rech J, Jaubert P, Rieth M (2007) Wiki-based stakeholder participation in requirements engineering. IEEE Softw 24(2):28–35. https://doi.org/10.1109/MS.2007.60

11. Knauss E, Brill O, Kitzmann I, Flohr T (3009) SmartWiki: support for high-quality requirements engineering in a collaborative setting. In: 2009 ICSE workshop on wikis for software engineering, pp 25–35, Vancouver, BC, Canada. https://doi.org/10.1109/WIKIS4SE. 2009.5069994
12. Abeti L, Ciancarini P, Moretti R (2009) Wiki-based requirements management for business process reengineering. In: 2009 ICSE workshop on wikis for software engineering, pp 14–24, Vancouver, BC, Canada. https://doi.org/10.1109/WIKIS4SE.2009.5069993
13. Kundi M, Chitchyan R (2017) Use case elicitation with framenet frames. In: 2017 IEEE 25th international requirements engineering conference workshops (REW), pp 224–231, Lisbon, Portugal. https://doi.org/10.1109/REW.2017.5310
14. Browning J, Adams R (2014) Doorstop: text-based requirements management using version control. J Softw Eng Appl 7(3):187–194. https://doi.org/10.4236/jsea.2014.73020
15. Chang C-H, Lu C-W, Chu W, Hsiung P-A, Chang D-M (2016) SysML-based requirement management to improve software development. Int J Softw Eng Knowl Eng 26(03):491–511. https://doi.org/10.1142/S0218194016500200
16. Siddiqui S, Bokhari M (2013) Needs, types and benefits of requirements management tools. Int J Trends Comput Sci 2(11):433–441. https://doi.org/10.1007/3-540-28244-0_14
17. Marques A, Ramalho F, Andrade W (2015) TRL: a traceability representation language. In: SAC '15 proceedings of the 30th annual ACM symposium on applied computing, pp 1358–1363, Salamanca, Spain. ACM. https://doi.org/10.1145/2695664.2695745
18. Maletic J, Collard M (2009) TQL: a query language to support traceability. In: 2009 ICSE workshop on traceability in emerging forms of software engineering, pp 16–20, ACM, Vancouver, BC, Canada. https://doi.org/10.1109/TEFSE.2009.5069577
19. Tausch N, Philippsen M, Adersberger J (2012) TracQL: a domain-specific language for traceability analysis. In: 2012 joint working IEEE/IFIP conference on software architecture and European conference on software architecture, pp 320–324, Helsinki, Finland. IEEE. https:// doi.org/10.1109/WICSA-ECSA.212.53
20. Femmer H, Fernández D, Wagner S, Eder S (2017) Rapid quality assurance with requirements smells. J Syst Softw 123:190–213. https://doi.org/10.1016/j.jss.2016.02.047
21. Pétin J-F, Evrot D, Morel G, Lamy P (2010) Combining SysML and formal methods for safety requirements verification. In: 22nd international conference on software & systems engineering and their applications. Paris, France. ACM. https://hal.archives-ouvertes.fr/hal-00344894
22. Chouali S, Hammad A (2011) Formal verification of components assembly based on SysML and interface automata. Innov Syst Softw Eng 7(4):265–274. https://doi.org/10.1007/s11334-011-0170-3
23. Insfrán E, Pastor O, Wieringa R (2002) Requirements engineering-based conceptual modelling. Requir Eng 7(2):61–72. https://doi.org/10.1007/s007660200005
24. Schätz B, Fleischmann A, Geisberger E, Pister M (2005) Model-based requirements engineering with AutoRAID. In: Proceedings of informatik 2005 workshop Modellbasierte Qualitätssicherung, pp 511–515, GI, Bonn (2005)
25. Zhu H, Lingzi J, Diaper D, Bai G (2002) Software requirements validation via task analysis. J Syst Softw 61(2):145–169. https://doi.org/10.1016/S0164-1212(01)00109-1
26. Gervasi V, Nuseibeh B (2002) Lightweight validation of natural language requirements. Softw Pract Exp 32(2):113–133. https://doi.org/10.1002/spe.430
27. International Institute of Business Analysis (2015) A guide to the business analysis bood of knowledge (BABOK Guide) version 3.0. International Institute of Business Analysis, Toronto, ON, Canada
28. Rich C, Sidner C (1998) COLLAGEN: a collaboration manager for software interface agents. Computational Models of Mixed-Initiative Interaction. Springer, Dordrecht, Netherlands, pp 149–184. https://doi.org/10.1007/978-94-017-1118-0_4
29. Fitzgerald B, Stol K-J (2017) Continuous software engineering: a roadmap and agenda. J Syst Softw 176–189. https://doi.org/10.1016/j.jss.2015.06.063
30. Bray T (ed) The javascript object notation (JSON) data interchange format. Internet Engineering Task Force, RFC 7158. https://buildbot.tools.ietf.org/html/rfc7158

31. Commonmark (2019) CommonMark parser and renderer in JavaScript. https://github.com/commonmark/commonmark.js
32. Chester D (2019) Query and manipulate JavaScript objects with JSONPath expressions. Robust JSONPath engine for Node.js. https://github.com/dchester/jsonpath
33. Poberezkin E (2019) ajv. https://github.com/epoberezkin/ajv
34. Mapbox node-sqlite3: Asynchronous, non-blocking SQLite3 bindings for Node.js. https://github.com/mapbox/node-sqlite3
35. Griesser T (2019) knex: a query builder for PostgreSQL, MySQL and SQLite3, designed to be flexible, portable, and fun to use. https://github.com/tgriesser/knex
36. Baszuro P, Swacha J (2014) Concept of meta-model describing collaboration in social networks. In: Marciniak A, Mikołaj M (eds) Social aspects of business informatics: concepts and applications, pp 23–32, Nakom, Poznań, Poland (2014)
37. Baszuro P (2019) Requirements management tool for continuous development. https://github.com/reqslang/reqslang

Information Management System for an Administrative Office with RFID-Tagged Classified Documents

Robert Waszkowski and Tadeusz Nowicki

Abstract The paper presents system architecture for the RFID-equipped restricted access administrative office. All the facilities in the administrative office are equipped with RFID readers placed in cabinets, desks, copiers and entrance sluices to allow immediately read of documents that are within their reach. Presented hardware and software components as well as business and simulation models are result of analytical work performed by the group of experts. The team was composed of specialist in different fields, such as document management, IT, data security, radio frequency identification, etc. In addition to the hardware and software architectures, an approach to system evaluation based on computer simulation is also presented.

Keywords RFID · System architecture · Security · Restricted access administrative office

1 Introduction

This paper presents design and development of an innovative electronic system for managing confidential document tracking. The research results are part of a scientific project partially supported by the National Centre for Research and Development under grant No. DOBR-BIO4/006/13143/2013.

The science project—entitled "Electronic system for life-cycle management of documents with various level of confidentiality"—focused on the use of the latest information and communication technologies related to Radio-Frequency Identification (RFID) and biometrics. The aim of completing the research project was to build and implement a prototype of a modern secret office designed to manage documents with various levels of confidentiality. The project covered all processes that take place in a secret office and is based on the use of devices equipped with RFID

R. Waszkowski · T. Nowicki (✉)
Faculty of Cybernetics, Military University of Technology, Warsaw, Poland
e-mail: tadeusz.nowicki@wat.edu.pl

R. Waszkowski
e-mail: robert.waszkowski@wat.edu.pl

© Springer Nature Switzerland AG 2020
A. Poniszewska-Marańda et al. (eds.), *Data-Centric Business and Applications*,
Lecture Notes on Data Engineering and Communications Technologies 40,
https://doi.org/10.1007/978-3-030-34706-2_3

transmitter/receiver modules. The focus of the research concerned not only documents and data carriers, but also all secret office equipment (cabinets, desk, copying machines and entry/exit control devices).

The integration of all elements in a single system, which is managed using dedicated software, increases the level of security and control over both classified and unclassified data.

The system allows tagging of electronic mediums as well as paper documents with RFID tags. Such marked documents, both sensitive and non-sensitive, electronic and paper, can be tracked. Tracking can take place both in the secret office and between security zones. The access to documents can be managed, logged and controlled, and documents can be protected from unauthorized copying.

The aim of the project was to prepare a prototype of the modern secret office, using the latest RFID technology, as well as to adapt office processes so as to work with documents with various confidentiality levels.

2 Objectives

The main objective of the project was to develop a modern system for electronic and paper document traceability based on RFID tags. The specific objectives of the project were:

- development of a real-time remote identification system for RFID tagged sensitive and non-sensitive media in both the workplace and storage facilities,
- development of an automatic inventory system for classified and unclassified documents arranged in stacks and contained in binders along with automatic detection of changes in their positions,
- development of a system to control the flow of media and classified and unclassified documents between the security zones, along with controlling document access permission,
- development of an electronic system for protection against unauthorized document relocation,
- automatic identification of media and documents in the storage area as well as in the workplace,
- development of technology to protect against multiple copying of the classified document,
- identification of the location of a single classified and unclassified document to the particular individual folder or volume.

3   RFID Technology

RFID is a technology of radio frequency identification. The idea of such a system is to store a certain amount of data in transceiver devices (RFID tags). Then the data is automatically read, at a convenient time and place, to achieve the desired result in a given application [1, 2].

The beginnings of radio frequency identification dates back to the forties of the twentieth century, when it appeared the device based on metal detectors. The first shop antitheft systems, based on the decoding of resonant circuit stickers or magnetoacoustic systems, which use magnetic shields, began to function from the sixties of the 20th century. Full radio identification appeared in the 1970s, and the first system introduced on the market was a Texas Instruments Tiris [1].

Today, the technology is growing fast and creates many new possibilities for potential users. An RFID System consists of two components: a tag placed on the identified object and reader, whose role is to read the data from the tag. Depending on its construction, the system allows for reading tags from up to a few inches or a few feet from the reader (Fig. 1). A positive aspect of this is that the transmission of data does not require the tag to be visible [1].

RFID tags are technologically advanced labels that contain electronic memory chip and antenna, through which the data is transmitted. RFID tags, depending on the nature of their application, are of different sizes and are made of various materials, mainly paper and plastic. Their construction depends on the requirements in the frequency range in which the device is to operate [1].

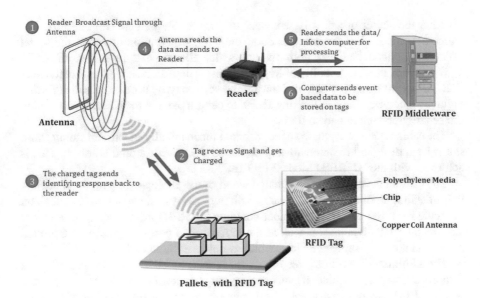

Fig. 1   RFID technology. *Source* http://www.reorientech.com/rfid.aspx

Tags can be classified according to their power source and memory type. Depending on the power source tags can be divided into Passive, Semi-Passive and Active. Depending on the memory type tags can be divided into Read Only and Read-Write.

Due to the technical implementation of RFID (encoding type, tag memory size, transmission speed, the ability to read multiple tags within range of the reader, etc.), there is a wide variety of RFID standards: Tiris, Unique, Q5, Hitag, Mifare, Icode, PJM [3, 4].

PJM (Phase Jitter Modulation) is a new form of encoding information in a radio wave through modulation of the phase. Although this technique has been developed recently, its essential capabilities allow achieving much better performance than any other previously known RFID standard. PJM technology enables efficient and error-free scanning a large number of tags, what sets it apart from the currently used RFID solutions, which disadvantages do not allow using them effectively for archiving and supervision of documentation [5, 6].

PJM allows faster reading a large number of tags in a shorter period of time, what plays a key role in identifying documents. Consequently, it was decided to use the PJM standard, which allows to read much larger amount of data stored in the tags and eliminates errors of identification in case of signal interference caused by the large number of stacked tags. All the other RFID technologies were not sufficient; therefore, the PJM technology was chosen to enable fulfilling project's goals.

4 Hardware

To enter the secret office, it is necessary to go through a sluice (Fig. 2, left). The sluice is composed of the entrance door (outside), which allows to enter the secret office from outside, and exit door (inside), which allow to exit the secret office.

It provides the functionality of RFID document detection as well as metal detection. All the information obtained during the scan may be used to initialize further archive procedures, such as limiting access to certain parts of the building, informing the security personnel and so forth.

The cabinet (Fig. 2, right) is an independent mechanically-electronic storage unit. It has been designed to automatically detect all documents stored within it. This is achieved with the HF RFID Mode 2 ISO 18000-3 tags.

The cabinet casing is made of coated wood or metal, depending on the model. A pair of sliding door is installed in each, with both a physical and a magnetic lock on it. Aside from the casing, the cabinet contains the RFID infrastructure (the reader and the internal antennas), as well as a control unit, a power supply unit, the main board and the electromagnetic lock which is unlocked with an identifier RFID tag.

The administrative office is also equipped with a copier. This device is based on a Canon 4225 copier machine. It synchronizes the process of copying of the document with the RFID tag readout of both the document and the RFID identifier of the user.

Fig. 2 The entrance sluice/body scanner (on the left) and the cabinet (on the right). *Source* Own elaboration

Based on the information read from the tags the centralized archive system permits the copier to create physical copies of only those scanned documents for which the user is authorized.

The copier includes a purposefully designed RFID antenna, a RFID reader and a server station modified to interface with the centralized archive system.

The tray reader was designed for the purpose of initial document registration. It automatically identifies, reads, writes and tracks the RFID-tagged documents. It is capable of reliable readout and writing of all RFID tags of the documents placed on the covered part of the tray.

Tightly packed stacks of overlapping or touching RFID tags are identified with 100% reliability. The reader operates instantly thanks to two reply channels and is capable of reliably identifying up to 60 documents placed in the tray.

The workstation for the restrictive access administrative office workers is equipped with desktop computer connected to all necessary peripheral devices including tray readers (Fig. 3).

Fig. 3 The restrictive access administrative office workstation. *Source* Own elaboration

5 Software

A combination of modern RFID technology, enabling rapid identification of a large number of objects in a short period of time, along with the appropriate computer system, that processes information collected by RFID devices, allows tracking and storing documents automatically. This applies to individual copies, as well as collections of documents (stored in folders, binders, or other office equipment adapted for this purpose). The use of a document management system is necessary to ensure the implementation of the objectives of the project. The system keeps track of document location changes (within the zone located in the secret office). It also provides information as to their current location and status.

In order to ensure communication between the document management system and RFID devices, it was necessary to use the appropriate system software that have been developed along with the RFID technology, enabling the transmission of information from RFID devices to application software for the collection and processing of the data. The number of utility programs will be used to build the system. CrossTalk AppCenter, developed by noFilis, is a modern platform for track & trace applications. The integrated object and event repository takes care of any data operation without touching the database itself. Event listeners are able to handle messages from various AutoID systems like passive RFID, active and passive RTLS, WIFI and GPS tags and many more. CrossTalk provides many adapters to exchange object and event data with backend or automation systems, including SAP (AII, IDoc, BAPI), Web-Services, databases, file interfaces and many more [7, 8].

The workflow in the designed solution is covered by the Aurea Business Process Management System. Aurea BPM is a multi-lingual and multifunctional tool for modelling, executing and optimizing business processes. It automates and improves dynamic, constantly changing business processes in heterogeneous, distributed, multi-language environments [9].

In the system for managing confidential document tracking, the JEE (Java Enterprise Edition) computing platform was used for building an application layer. As a database the Oracle Relational Database Management System have been used. It provides safe and reliable data storage for the tracked documents.

The system (Fig. 4) allows the current and historical control and analysis of media and classified and unclassified documents flow, both within the security zone of secret office as well as any changes resulting from a change of the zone (entry/exit). Any object relocation is associated with a person who made this relocation. Depending on the permission of such persons, the appropriate response of the system is taken. If the person has no access right to the document, an alert is automatically generated. The system also uses biometric technology, which gives the opportunity to "bind" the user with its record in the system. User permissions to the specific type of documents is being determined from the permissions management system.

In the case of unauthorized document relocation, depending on the assumptions, the system responds automatically. In addition, the body scanner is used for this purpose. It is installed at the border zone (entry/exit) and detects both metallic and non-metallic items, including RFID tags and other metallic items that could be used to hide tags in. The use of RFID tags, the quality readers for reading and automatic identification of users, gives a full view of the displacement of the media or documents.

All copiers used in the secret office are equipped with an RFID reader, which gives you the ability to monitor and manage the creation of photocopies of documents

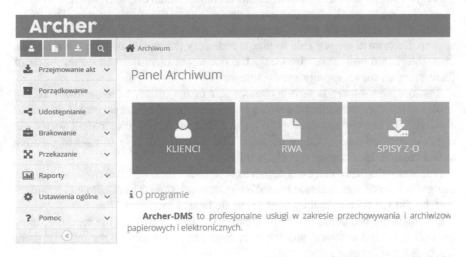

Fig. 4 The restrictive access administrative office management system. *Source* Own elaboration

(classified and unclassified). Depending on user permissions and the type of the document, the system will either allow or deny copying. All photocopies are printed on a paper with assembled RFID tag, which then is recorded in the system as a subsequent copy of the document. It allows the system to manage all copies of the document.

6 Business Processes

As part of the project, for the document processing purpose, it was necessary to integrate with devices equipped with RFID mechanism for document and user identification. The implemented rfiDoc system is aimed not only at collecting events from RFID devices, but also at controlling the flow of the business processes on the basis of such events. The communication between the rfiDoc system and RFID devices is mutual. A dedicated module responsible for the aforementioned communication was implemented in the rfiDoc system.

As part of the communication between the devices using the RFID mechanism, the following elements were distinguished:

Cabinets—there are two types of cabinets available in the office, i.e. cabinets with sliding door and cabinets with double-leaf doors. All of the above-mentioned types of cabinets have the RFID mechanism installed to control access and monitor their contents. The access control is executed on the basis of the map of privileges sent by the dedicated module of the rfiDoc system to every cabinet agent independently. The map of privileges is a list of RFID-access key tags. On the other hand, the content of the cabinets is automatically sent by the cabinet agent to the dedicated module of the rfiDoc system each time the door is closed (Fig. 5).

Trays—the Office clerk also has the RFID readers at its disposal, which are called trays allowing to quick identification of documents in the rfiDoc system. The RFID document tag read by the tray is directly sent to the dedicated module of the rfiDoc system. Such trays are not able to directly identify the person making the read-out, thus, the rfiDoc system receives only the document identifier and on the basis of them searches for an appropriate document.

Photocopiers—there is also a photocopier in the office, which allows making copies of the documents that the office may access based on the access rights vested therewith. In case of any attempt to copy a document, the printer shall verify the rights in the dedicated module of the rfiDoc system and only in case of positive authentication, the copying process is initiated.

Tunnels—the office clerk has access to various devices, which allow searching for a large number of documents in the rfiDoc system at a time. The document identifiers read by the tunnel are directly sent to the dedicated module of the rfiDoc system. The tunnels are not able to directly identify the person making the read-out, thus, the rfiDoc system receives only the document identifiers and on the basis of them searches for appropriate documents.

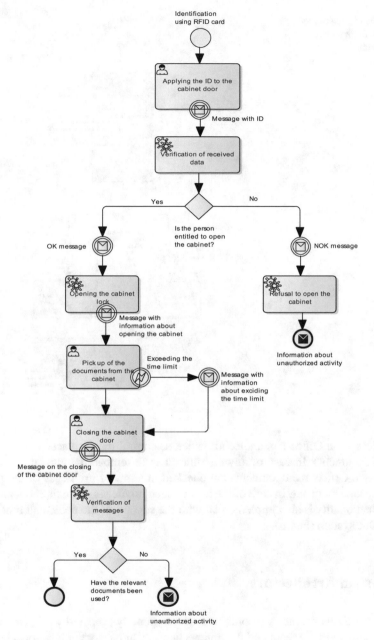

Fig. 5 Cabinet operating diagram. *Source* Own elaboration

Fig. 6 Sluice operating diagram. *Source* Own elaboration

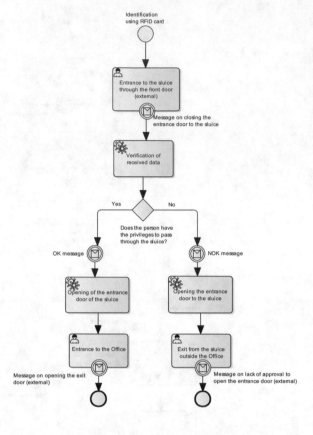

Sluices—the Office has a special device to control the displacement of RFID-tagged documents. In case of any attempt made to remove the tagged documents outside the Office, the documents are checked in terms of privileges of the person trying to take them out. In case of the lack of such privileges, the exit doors are shut. This functionality is also implemented with the support of the dedicated module of the rfiDoc system (Fig. 6).

7 System Architecture

The use of RFID technology combined with a properly prepared workplace allows full identification of media and documents not only at its storage locations, but also at all the user workstations. It is assumed that all workplaces in the security zone will be equipped with properly designed furniture (fitted with suitable reader for document identification) [10–16].

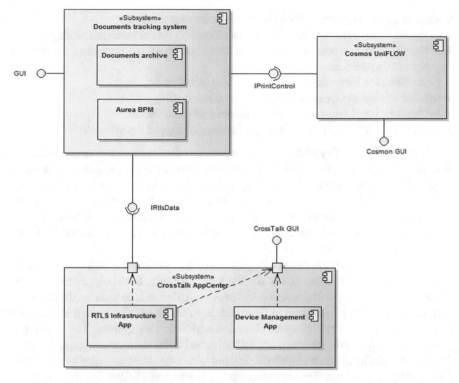

Fig. 7 System architecture. *Source* Own elaboration

The logical architecture of the system is shown in Fig. 7. The diagram shows the external interfaces of the document lifecycle management system, which will cooperate with CrossTalk's AppCenter and Cosmos UniFLOW. Communication will occur by the programming interfaces—Web services [17, 18]. Document Management System will also be equipped with the graphical user interface (GUI), available from a Web browser.

8 Experimental Evaluation of System Reliability

Reliability model for document management system in the environment located in one room and made up of many different types of elements, including RFID antennas, metal and wooden cabinets with drawers, cabinets, photocopiers, recording desks, locks, badges and a large number of documents marked with RFID identifiers was developed. Each component was classified in terms of the reliability. For most of them individual reliability models were developed.

Elements of these reliability models connected with the whole system includes:

- elements of the remote identification of documents marked,
- elements of the remote identification of electronic media (external storage),
- elements of the remote registration of the position of the document marked or other media positions,
- procedures for inventory of documents and other media,
- models of change of position papers and other media,
- procedures for sharing documents marked and other media,
- procedures for selective copying and printing of tagged documents.

The level of reliability of the document management system and media is dependent on the efficiency of the system's hardware components state, including a set of RFID antennas, and the reliability of system software. In the process of reliability investigation of the software, connected with document management system, the knowledge of the software component structure (modular structure) is used. Indicator of system software reliability will be a function of this structure and values of the reliability of software components.

Reliability states of individual components of the system were defined (hardware and software) and the same states for the whole system. Similarly, for individual documents and their packages suitable states were defined, as well as the relationship between different elements in the reliability model, and the system in the form of multi-dimensional reliability structural function of this system. The stochastic process associated with the transition of individual documents or packages between different reliability states was defined. It determines the characteristics of reliability for the workflow system. These characteristics are the basis to develop the reliability ensuring plan required for modern technical systems. Under this reliability ensuring plan following elements were developed:

- measurement reliability of the system and its separate elements,
- measure software reliability and its constituent components,
- levels of limit values for the reliability of the system and its separate elements,
- list of researches necessary to investigate hardware, procedures and software components,
- methodology for testing various hardware components of the system,
- methodology for testing individual components of system software,
- methodology for the study of individual procedures,
- the form of the final report of reliability tests.

The whole methodology is a mature analysis of the reliability of the system. It provides a basis for testing reliability and testing efficiency in the subsequent steps of the project.

Reliability testing document and media (external memory) management system and a study of its subsystems are divided into two stages:

- examination of the system and its separate parts beyond the prototype of office space,
- examination of the system and its separate parts in the prototype of office space.

Using the previously developed methodologies and reliability characteristics of the study it is subject to:

- remote identification documents and media, unclassified and classified, storage places and work in real time,
- procedures for automated inventory of documents and media, unclassified and classified placed in stacks,
- the automatic detection of changes in the position of marked documents and media,
- flow control of unclassified and classified documents and media between security zones,
- checking and controls procedures of persons authorization to classified documents,
- a system of electronic documents and media protection against unauthorized movement and identification documents and media on the workplace,
- method to prevent multiple copy a document and print a limited number of copies,
- identification of the position of a single document and media from the specified accuracy,
- proper functioning of the special entire gate.

These studies were carried out both in relation to the entire document management system as well as its individual parts.

The result was developed for each of the two test stages:

- final reliability values for system and its separate parts,
- final reliability values for software and its separate components,
- recommendations arising from the study of reliability of the system and its separate parts,
- final report of reliability tests.

All the hardware was tested. The system cooperates with many hardware elements such as cabinets, copiers, trays, tunnels, etc. An example of the testing procedure for the body scanner is presented below.

It was assumed that the RFID CCS D2 body scanner is irreparable. Defective scanner is exchanged for a new, uniform, in the sense of reliability, with the scanner used so far. Since the commission does not have data from the manufacturer about the reliability of the RFS CC2 body scanner, it is based on the results of reliability tests of similar systems that the scanner lifetime is variable random X with an exponential distribution with the parameter λ.

In this case, the reliability indicators assume courses of reliability function R(t) and intensity function $\lambda(t)$. Figure 8 presents the courses of these reliability characteristics of the RFID CCS D2 body scanner.

9 Experimental Evaluation of System Efficiency

All relevant methods, procedures, techniques, etc. in document management system must be examined in terms of the efficiency of their use. It is very important to

Fig. 8 Reliability function for the body scanner RFID CCS D2: **a** R(t), for t ≥ 0; **b** λ(t), for t ≥ 0. *Source* Own elaboration

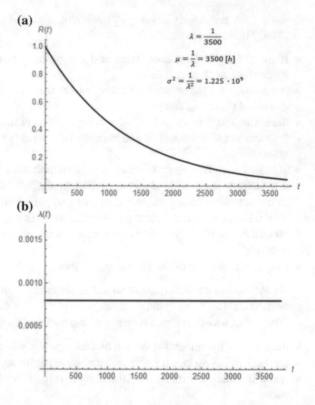

examine the capacity of the system or its separate components. In this case, as in reliability studies of document management system, we study system and its separate subsystems efficiency in two stages:

- examination of the system and its separate parts beyond the prototype of office space,
- examination of the system and its separate parts in the prototype of office space.

Research related to the term efficiency of the system may depend on a number of important factors affecting the efficiency of the system:

- size of sets of documents and media marked by RFID in selected procedures of their circulation,
- time duration of set of documents and media marked by RFID identification in selected procedures of their circulation,
- the significance of the impact of typical and random changes in system conditions to correct his work.

Additionally, there were determined characteristics related to the load of the system software.

It was decided to use an innovative method to test the reliability and efficiency of the system. A simulator imitating document flow events was created, in this case

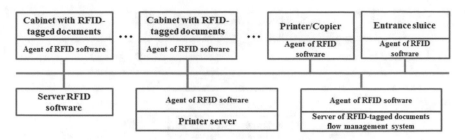

Fig. 9 System supporting the RFID-tagged document flow. *Source* Own elaboration

generating signals that change the location of documents with RFID tags in the office system. Event simulators simulating the functioning of technical systems have been described in many papers and monographs. In normal mode, the RFID software server is only used to configure a scanner, which defines the devices that are part of the company's office.

In turn, the existence of documents with RFID tags is recorded because RFID antennas send messages to RFID software agents and these are then transmitted by the RFID software agent to the Document Management Server. Messages are XML strings. Thus, the idea of building a simulator of XML messages imitating the existence or non-existence and movement of documents in the system.

For this purpose, a replacement of the original work environment of the office was shown, as shown in Fig. 9. In this case, a series of event scenarios would be required for messages sent between RFID software agents from external devices to the RFID server of the system. In addition, you will have to track the response of the document flow management server to the occurrences.

It was decided to use an innovative method to test the reliability and efficiency of the system. A simulator imitating document flow events was created, in this case generating signals that change the location of documents with RFID tags in the office system. Event simulators simulating the functioning of technical systems have been described, inter alia, in the model sense in monographs [19] and [20], and in the sense of implementation in monograph [21] and [22]. In normal mode, the RFID software server is only used to configure a scanner, which defines the devices that are part of the company's office.

In turn, the existence of documents with RFID tags is recorded because RFID antennas send messages to RFID software agents and these are then transmitted by the RFID software agent to the Document Management Server. Messages are XML strings. Thus, the idea of building a simulator of XML messages imitating the existence or non-existence and movement of documents in the system. The environment of the RFID tag assisted document flow system was transformed into the artificial environment illustrated in Fig. 10.

This makes it possible to prioritize RFID status papers in the office with their locations. This way, in the finite, soon to be able to study the reliability characteristics And the efficiency of the office system both in terms of the application software of

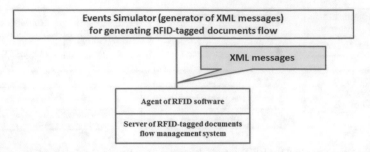

Fig. 10 Process related to the storage of the RFID-tagged documents. *Source* Own elaboration

the Document Management Server as well as the correct operation of records In the database. Without this modification, the study of characteristics would be extremely long.

The simulator should enable testing and testing of the performance and reliability of the RFID tag flow management system in a dedicated, complex system. The simulator is supposed to generate events by imitating activities carried out in a secret office equipped with specialized software. Currently, the following events are supported:

- Opening the front door—when opening the lock door, the device sends information about this event along with information about who did it and list of all RFID tags currently on the device,
- Closing the entrance door—When closing the lock outlet door, the device sends information about this event along with who it was and the list of all RFID tags currently on the device,
- Document Location—At the document location, the device sends information about this event along with a list of all RFID tags currently on the device,
- Document detection—When a document is detected, the device sends information about this event along with a list of all RFID tags currently on the device.

The event-based scenario simulator constructs XML messages accepted by the server module and then sends them via http. The simulator has a graphical interface made using the JavaFX library and requires Java 8 virtual machine software (1.8). The simulator window is shown in Fig. 11.

Before you run the simulation, you may need to update your RFID agent data. This is done by a special button that initiates the required data based on a defined scenario. During the normal operation of the simulator, the events generated by the nodes (generators) are stored in the event log (Fig. 12). In addition, they are obviously sent to the RFID agent service.

After the simulation is complete, you can preview the report of the experiment (Fig. 13).

Thanks to this simulator, the annual operation of the system was simulated in a short time. Several software errors found. It is an excellent method of studying the properties of complex software in systems in which human activity is characterized by relatively low intensity.

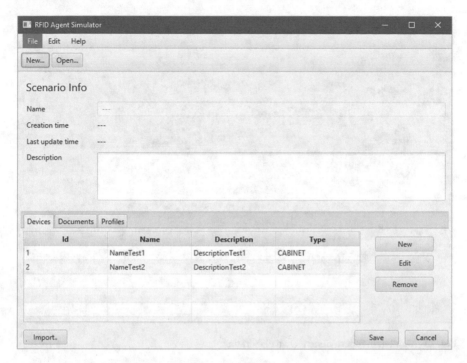

Fig. 11 Main simulator window. *Source* Own elaboration

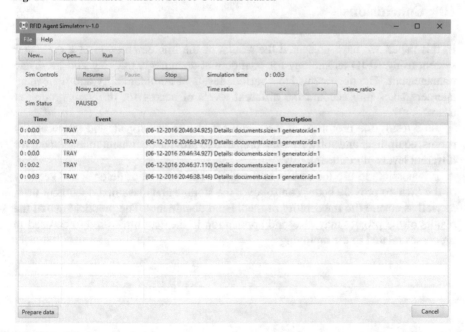

Fig. 12 End of experiment. *Source* Own elaboration

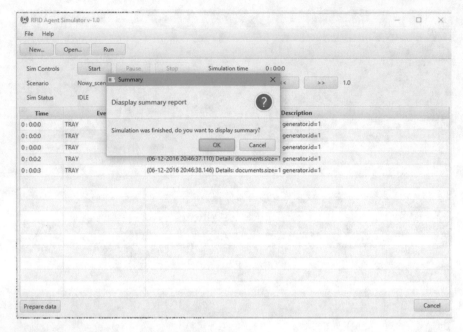

Fig. 13 Simulation report. *Source* Own elaboration

10 Conclusions

In this paper, we have presented the results of the implementation of a computer system using RFID technology for confidential documents tracking and life-cycle management. The documents have various levels of confidentiality and their management takes into account the different levels of access for different groups of users.

As a result, the prototype of the secret office was developed, and all necessary rooms, equipment and software were prepared to provide document management for different levels of confidentiality.

It is possible to use the results of the project in the area of defense and security of the State to provide better control over the storage and classified document flow, as well as ensure the traceability of media and documents. The practical use of the results of the project may also find application in the economic area to be used in processes related to the archiving.

References

1. Finkenzeller K (2003) RFID handbook: fundamentals and applications in contactless smart cards and identification, 2nd edn. Wiley
2. Cole PH, Ranasinghe DC (2008) Networked RFID systems and lightweight cryptography. Springer
3. Zhang Y, Yang LT, Chen J (2009) RFID and sensor networks architectures, protocols, security and integrations. CRC Press
4. Paret D (2009) RFID at ultra and super high frequencies. Theory and application. Wiley
5. Bolic M, Simplot-Ryl D, Stojmenovic I (2010) RFID systems research trends and challenges. Wiley
6. Miles SB, Sarma SE, Williams JR (2008) RFID technology and applications. Cambridge University Press
7. noFilis, CrossTalk AppCenter 3.0 Installation and Administration Guide
8. Canon UniFLOW documentation. www.canon.com
9. Aurea BPM system documentation. aurea-bpm.com
10. Kiedrowicz M, Nowicki T, Waszkowski R, Wesolowski Z, Worwa K (2016) Method for assessing software reliability of the document management system using the RFID technology. In: Mastorakis N, Mladenov V, Bulucea A (eds) 20th international conference on circuits, systems, communications and computers, vol 76. MATEC Web of Conferences, Cedex A: E D P Sciences
11. Kiedrowicz M, Nowicki T, Waszkowski R, Wesolowski Z, Worwa K (2016) Optimization of the document placement in the RFID cabinet. In: Mastorakis N, Mladenov V, Bulucea A (eds) 20th international conference on circuits, systems, communications and computers, vol 76. MATEC Web of Conferences, Cedex A: E D P Sciences
12. Kiedrowicz M, Nowicki T, Waszkowski R, Wesolowski Z, Worwa K (2016) Software simulator for property investigation of document management system with RFID tags. In: Mastorakis N, Mladenov V, Bulucea A (eds) 20th international conference on circuits, systems, communications and computers, vol 76. MATEC Web of Conferences, Cedex A: E D P Sciences
13. Nowicki T, Pytlak R, Waszkowski R, Bertrandt J (2013) The method for finding optimal sanitary inspectors schedules (in English). Ann Nutr Metab 63:1027
14. Nowicki T, Pytlak R, Waszkowski R, Zawadzki T, Migdal W, Bertrandt J (2013) Creating and calibrating models of food-borne epidemics (in English). Ann Nutr Metab **63**:1027
15. Waszkowski R, Agata C, Kiedrowicz M, Nowicki T, Wesolowski Z, Worwa K (2016) Data flow between RFID devices in a modern restricted access administrative office. In: Mastorakis N, Mladenov V, Bulucea A (eds) 20th international conference on circuits, systems, communications and computers, vol 76. MATEC Web of Conferences, Cedex A: E D P Sciences
16. Waszkowski R, Kiedrowicz M, Nowicki T, Wesolowski Z, Worwa K (2016) Business processes in the RFID-equipped restricted access administrative office. In: Mastorakis N, Mladenov V, Bulucea A (eds) 20th international conference on circuits, systems, communications and computers, vol 76. MATEC Web of Conferences, Cedex A: E D P Sciences
17. Braude EJ, Bernstein ME (2011) Software engineering: modern approaches. Wiley
18. Larman C (2012) Applying UML and patterns: an introduction to object-oriented analysis and design and iterative development. 3/e, Pearson Education India
19. Ross SM (2006) Simulation. Elsevier Inc
20. Sinclair B (2004) Simulation of computer systems and computer networks: a process-oriented approach. University Press, Cambridge, UK
21. Fishman GS (2001) Discrete event simulation. Modeling, programming and analysis. Springer, New York
22. Abu-Taieh EM, El Sheikh AAR (2010) Handbook of research on discrete event simulation environments: technologies and applications. IGI Global, Hershey, New York

Framework for Processing Behavioral Business Rules Written in a Controlled Natural Language

Bogumiła Hnatkowska and Martyna Litkowska

Abstract Business rules are the basis of the business logic of the most information systems. They are considered to be the first citizens of the requirements world and are the key element of business and technology models. It is highly recommended the rules to be written in a natural language, which is understandable for all interested parties. The paper presents the actual state of a framework for business rules processing. The framework is capable of serving business rules written in a controlled language (which syntax was inspired with SBVR SE) assuming that the source code is properly instrumented. Unfortunately, there exists a very limited number of solutions enabling processing business rules expressed that way. The proof-of-concept implementation proved the correctness and usefulness of the proposed approach.

1 Introduction

Business rules are the basis of the business logic of the most information systems. They are considered to be the first citizens of the requirements world [1] and are the key element of business and technology models. They are vulnerable to changes and therefore are a sensitivity point of many applications. In most cases, a change in a business rule forces a change in the business logic source code with all the consequences—the new source code has to be built, deployed, and a new instance of the software system has to be run. The application of special solutions can minimize the above mentioned negative effect of a business rule change—they include, e.g., specialized integration layers, business rules engines like JBoss Drools or both [2, 3]. A business rule can be changed internally within a business rule engine, transparently to the application itself.

B. Hnatkowska (✉) · M. Litkowska
Faculty of Science and Management, Wrocław University of Science and Technology, Wyb. Wyspiańskiego 27, 50-370 Wrocław, Poland
e-mail: bogumila.hnatkowska@pwr.edu.pl

M. Litkowska
e-mail: martyna.litkowska@gmail.com

© Springer Nature Switzerland AG 2020
A. Poniszewska-Marańda et al. (eds.), *Data-Centric Business and Applications*, Lecture Notes on Data Engineering and Communications Technologies 40, https://doi.org/10.1007/978-3-030-34706-2_4

Business rules can be expressed in different syntaxes, from more programming oriented, e.g. OCL [4], JDrools DSL [5], to more users oriented, e.g. SBVR Structured English (SBVR SE) or RuleSpeak. SBVR SE is a controlled natural language introduced in SBVR—an OMG standard "for the exchange of business vocabularies and business rules among organizations and between software tools" [6]. According to the Business Rules Manifesto [1], the natural language should be preferred as it is easier to understand by all interested parties. Unfortunately, the existing solutions to business rule representation used in commercial applications do not support processing of natural language. The tools require its users to be familiar with a specific formal notation, which can be a mixture of a declarative language (see e.g. "when" part in a rule definition in DRL language used by Drools engine), and operative one (e.g. "then" part). That significantly limits the solution accessibility, e.g. for business analysts without a technical background.

Business rules are typically classified into two main categories [6]: static (also called definitional), and behavioral (also called operational). Static rules are simpler to be served by tools as they relate to entities, their properties, and relationships among them. Very often they represent integration rules (e.g. invariants, multiplicity constraints) which can be checked at different application layers, e.g. by presentation layer or a database engine. Most of the modern programming languages offer a declarative way (annotations) for constraints definition (see e.g. [7, 8]). There also exist several solutions for structural rule maintenance, e.g. [9, 10].

Behavioral business rules indicate something people or organizations are either obliged to do (an obligation), or prohibited from doing (a prohibition) [6]. Researchers do not so often address them. One example is [11], but here, the rules are expressed in OCL. The other is [5], where post-conditions expressed in OCL or Drools DSL are translated to Java annotations. In both cases, behavioral business rules are expressed in a formal language which can't be easily understood by all interested parties. Theoretically, these problems could be addressed, e.g., by application of solutions which are able to translate statements written in unstructured natural language to more formal representations like SBVR SE or OCL. Such a method is described in [12, 13]. Unfortunately, the method is still under development, and now the prototype tool [14] which is able to process only simple and correct English sentences. Another research addressing, among others, the processing of behavioral business rules is [3], in which a concept of integration layer which allows splitting business logic from business rules is presented. The integration layer is built upon Aspect Oriented Programming (AOP) paradigm. On the base of DSL statements, linking the integration layer linking the business logic and business rules, aspects are generated. However, in this solution business rules are written in DRL language (Drools engine) which requires advanced skills and full knowledge about the way information model is implemented.

The limitations mentioned above to the behavioral business rules management resulting from the way they are represented in a software system were addressed by a framework, which allows business rules to be expressed in a semi-natural langue with an SBVR SE-like syntax [15]. The paper presents its actual state. It is the next step in achieving a solution mature enough to be applied to a real software system.

The framework architecture and the extensions are described in Sect. 2. An example of framework usage is shown in Sect. 3. The critical analysis of the proposed solution is given in Sect. 4.

2 Framework Description

The framework is a "skeleton of interlinked items which supports a particular approach to a specific objective and serves as a guide that can be modified as required by adding or deleting items" [16]. In the context of the paper, the interlinked items are framework components, and the specific objective is an automatic processing (possibility to introduce a change) of business rules expressed in a natural language without the necessity of the application restart.

2.1 Business Vocabulary and Behavioral Business Rules

Definition of any business rule requires the existence of a vocabulary containing terms and facts. Business rules are built on facts, and facts are built on concepts as expressed by terms [1]. To support a user with a vocabulary definition, a vocabulary editor was constructed with the use of Xtext tool [17]. The tool automatically provides an appropriate parser (based on of delivered grammar) and generates an editor with colored syntax and contextual hints. The vocabulary editor works, e.g. as an Eclipse plugin. An example of the vocabulary definition is presented in Fig. 1.

The example presented above introduces three terms (*customer, order, car*), and two binary facts (at that moment, n-ary facts are not supported). The vocabulary content could be represented graphically, e.g. with the use of the UML class diagram, on which terms would be classes or their attributes, and facts—associations or generalizations. It should be mentioned that the vocabulary must be consistent with the source code (so-called model implementation). For the example given in Fig. 1, it is expected that there exist implementation of *Order, Customer*, and *Car* classes. Otherwise the framework won't be able to work correctly. However, the problem can be addressed by the automatic vocabulary generation from the selected modules (packages) from the source code.

```
package pl.edu.pwr.example {
  Term: customer Description:
      'Driver that bought something.'
  Term: order
  Term: car
  Fact: customer places order
  Fact: customer rents car
}
```

Fig. 1 Example of business vocabulary

The previous framework implementation [15] supported two types of behavioral business rules from The Business Rules Group classification [18]:

1. Pre/post conditions for actions: they check if a specific condition is true before/after a specific action. Such rules represent part of a contract. If the rule is broken, an exception is thrown;
2. Action enablers: they run associated actions if a condition is true; they are typical "when condition then action" rules expressed in natural language.

The proposed syntax to express business rules tried to mimic obligation ("it is obligatory …") and prohibition ("it is prohibited …") statements from SBVR classification [6]. It should be noted that the rules don't need to follow all grammar rules (see, e.g., missing articles). The exact moment when the rule will be checked is specified either directly (after/before) or indirectly (if, when—the same meaning as for before). Specification of the moment a rule is expected to be validated is an extension to the original SBVR SE syntax. Each rule contains a condition to be checked. The first version of the framework was able to verify two primary cases:

1. Terms comparison: two terms can be compared with the use of a relational operator, e.g. "If a driver rents a car, then it is obligatory that the age of the driver is at least minimumAllowedDriverAge of car" (terms to be compared: "the age of the driver" and "minimumAllowedDriverAge of car").
2. Comparison of a term and a literal value; the value must be an integer, e.g. "If a driver rents a car, then it is obligatory that the age of driver is at least 18 (elements to be compared: "the age of driver" and 18).

Definition of behavioral business rules is supported with another editor built with Xtext based on of proposed grammar, which is used later for syntax hints—see an example in Fig. 2. The editor is delivered as a web application. A simplified example of a grammar rule (in BNF notation) representing a business rule is given below:

```
<RuleName> <ActionPlace> <Clause>, (then)?
<ModalOperator>
(<BinaryTestClause> | <InvocationClause>)
```

Fig. 2 Business rules editor in action

where:

- *RuleName*—rule name, e.g. "Rule MinimalLevel is"
- *ActionPlace*—where to check the rule (before action or after it)
- *Clause*—action to be run
- *ModalOperator*—obligation or prohibition statement
- *BinaryTestCase*—condition type to be checked
- *InvocationClause*—action to be run for action enabler rule

2.2 Framework Architecture

Figure 3 shows a schema of business rules processing within the framework. In the first stage, the business vocabulary is defined in the vocabulary editor. In the second stage, business rules are formulated in the business rules editor. After that, the rules are translated to the form accepted by JBoss Drools rule engine [19], to be more specific, rules are translated to the MVEL dialect supported by Drools. Translated results are kept internally in a local cache (*RuleRepository* class), which can establish a session with Drools, set up the proper context, and evaluate selected business rules.

An example of a rule expressed in natural language (Rule CustomerHasMoney is: If customer rents car, then it is obligatory that the money of customer is at least the cost of the car) translation to source code in MVEL dialect is shown in Fig. 4.

The framework assumes the source code be appropriately instrumented. All the methods, implementing behaviors which change the state of the system, and are potentially involved in business rules, should be marked with @BusinessRule annotation, also describing method parameters. This information is used further to connect a specific business rule with a part of source code in a transparent way for the application itself with the use of Aspect Oriented Programming features. Functional architecture of the framework is presented in Fig. 5. The processes to serve a business rule is realized by different components, i.e. Vocabulary editor, Business Rule Editor, Business Rules Processor, and Business Rule API.

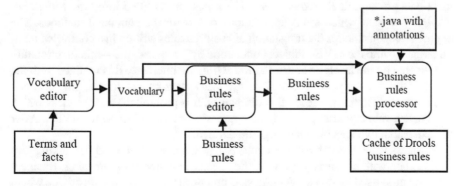

Fig. 3 Schema for business rules processing in the framework ([15])

```
PackageDescr pkg =
  DescrFactory.newPackage().name("pl.edu.pwr.gawedat")
    .attribute("dialect").value("mvel").end()
    .newImport().target("example.Car").end()
    .newImport().target("example.Customer").end()
    .newRule().name("CustomerHasMoney")
       .lhs()
       .pattern("RuleInfo").id("$ruleInfo", false)
       .constraint("businessAction=='customer rents car'")
       .constraint("when == 'before' ").end()
       .pattern("Car").id("$car", false).end()
       .pattern("Customer").id("$customer", false).end()
       .eval()
         .constraint("! ($customer.money >= $car.cost)")
       .end()
     .end()
    .rhs("$ruleInfo.reportBroken('CustomerHasMoney')")
    .end()
  .getDescr();
```

Fig. 4 Example of business rule translation to JDrool's format

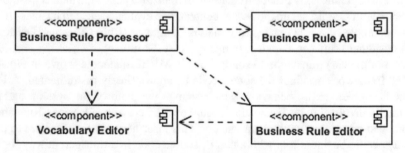

Fig. 5 Functional architecture of the business processing framework ([15])

The core element of the architecture is the Business Rules Processor. Main elements of this component (after the extensions) are shown in Fig. 6. Business rules meta-data are kept in *RuleRepository*. The repository is filled in by the *RuleBean-Scanner*—a class which scans application context for the annotated methods. The engine is responsible for the translation of business rules written in a controlled natural language into the MVEL dialect understood by Drools engine—this functionality is delivered by the implementation of *RuleMapping* interface. It is realized with the usage of the strategy pattern. The *StrategyHolder* class knows the mapping between a given modal expression (e.g., It is obligatory, It is prohibited) and the class which can parse a rule containing the modal expression and translate in to MVEL. After the extension, two specific mappings are defined.

Business Rule API delivers elements, e.g. java annotations, used for linking java source code with business logic with the Aspect-Oriented Programming. Business rules are processed by one aspect (AspectJ and Spring AOP [20]) with advice which

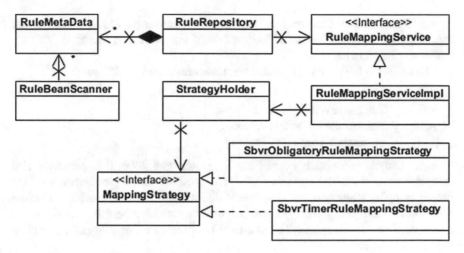

Fig. 6 Core classes of business rule processor

is run around every method with @BusinessRule annotation. It is responsible for filtering the proper business rules for the specific method on the base of annotation parameters, and their evaluation. If the business rule is not met an exception is thrown. Dependently on the rule type, sometimes a specific action is run as a result of rule evaluation.

2.3 Framework Extensions

2.3.1 Timers Support

The list of business rule types served by the framework was extended with a new type—timer. A timer is a type of action assertion which tests, enables (or disables), or creates (or deletes) if a specified threshold has been satisfied" [18]. The proposed grammar for timers is a little bit limited. One can define the test part of the object controlled by the rule (e.g. "if tank of car is less than 15.5"), and then enable/disable existing rules (e.g. "customer can rent car"). Two separate grammar rules were proposed to support the new business rule. The first is shorter and consistent with other rule types:

<RuleName> (If)
(<BinaryTestClauseWithIntValue>
  | <BinaryTestClauseWithDoubleValue>
) (, then)
(<CanClause> | <CannotClause>)

e.g. "Rule MinTankCanRent1 is: if tank of car is greater than 15.5, then customer can rent car" or "Rule MinTankCanRent2 is: if tank of car is less than 30, then customer can't rent car".

The second is SBVR like [6] and it resembles restricted permissions:

<RuleName> (Only if)
(<BinaryTestClauseWithIntValue>
 | <BinaryTestClauseWithDoubleValue>
) (, it is permitted that) (<Clause>)

e.g., "Only if production year of car is greater than 2008, it is permitted that customer rents car". The servicing of timers required the implementation of the proper mapping strategy in the Business Rule Processor. That was achieved by the *SbvrTimerRuleMappingStrategy* class, responsible, among others, for translation of a Rule object to Drools syntax. The extension confirmed the flexibility of the original engine architecture.

2.3.2   Double Values Support

Business rules can refer to numbers, especially their condition part. Up to now only integer values were supported. One of the framework extensions addresses this problem allowing usage of double values as well (both positive and negative). Exemplary usages are presented below:

- cubicCapacity of car is at least 1.8
- accountBalance of customer is greater than –9.90.

2.3.3   Direct Representation of Attributes and Roles

A (binary) fact links two terms (they could be also roles) with a verb, e.g. "customer has a name", "car complies with a model", what can be defined graphically as it is shown in Fig. 7.

In that example, "name" is a feature of every "customer", why "model" is a role played by *CarModel* class instance in the relation represented by "complies with" fact. Terms and facts limit the way a business rule can be expressed—any navigational expression must be consistent with them. In consequence, the vocabulary has to contain not only classes (*Customer*) but also roles and features, and all the facts linking them.

The extension to the existing framework allows expressing the same fact in a simplified manner. The vocabulary part was extended with attributes section—see an example below—in which for the "customer" the "name" and "model" were defined:

Fig. 7 Facts graphical
representation

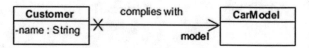

Term: customer
  Description: 'Someone who rents a car'
  Attributes: 'name', 'model'

Attributes include both: features of a specific class (direct attributes like, e.g., name), and roles visible from the term perspective (e.g. model). Such a definition makes possible navigation to the entitys attributes (or roles) with "of" expression, e.g. "name of customer", "model of car". That simplifies later business rule definition.

2.3.4 Synonyms

To widen the solution flexibility, the concept of terms synonyms was addressed. Now, a specific term can have multiple synonyms which can be used in every place the term is expected (it is assumed synonyms are unique in the vocabulary):

Term: customer
  Description: 'Someone who rents a car'
  Synonyms: 'client'

In the example presented above, two synonyms for "customer" were introduced ("client" and "person"). Each of them can be used interchangeably instead of the customer in the business rule statement, e.g. two rules given below have exactly the same meaning and effects:

- Rule1 is: if tank of car is less than 15.5, then customer can rent car
- Rule2 is: if tank of car is less than 15.5, then client can rent car.

Servicing of synonyms was also included in Business Rules Editor which, after extension, suggest proper ones—see Fig. 2.
It is assumed that synonyms, like terms, are unique to avoid ambiguity.

2.3.5 The Canonical Representation of Business Rules

Implementation of the above-mentioned extensions, especially synonyms, required changes in internal business rules representation. Authors introduced so-called the canonical representation of business rules, which splits a business rule into three parts: left (subject), middle (action), and right (complement). This decision is reflected in the syntax of annotations for the source code (see Fig. 8).
The methods in the source code for which business rules could be applied have to be annotated with @BusinessRule annotation. Before the change, this annotation has the "value" attribute set to one of the facts from the business vocabulary, e.g. "customer rents car". As business rules are built upon facts, for the method with such annotation all business rules built upon the fact ("customer rents car") were checked what wasn't very effective. The problem with the approach is that the "value" attribute is a text, not very useful for further filtering. It is why the annotation structure was

```
@BusinessRule(
  value = "customer rents car",
  action = "rent",
  subjects = { "customer", "client" },
  complements = { "car" },
  lookups = {
  @BusinessTermLookup(name = "customer",
      path = "arg[0].customer"),
  @BusinessTermLookup(name = "client",
      path = "arg[0].customer"),
  @BusinessTermLookup(name = "car", path = "arg[0].car")}
)
public Car createRental(Rental rental) { ... }
```

Fig. 8 Example of annotated service

```
String businessAction =
  rule.getAction().getLeftSide().getName() + " "
+ rule.getAction().getAction() + " "
+ rule.getAction().getRightSide().getName();
String place = resolveWhen(rule.getPlace());

CEDescrBuilder<RuleDescrBuilder, AndDescr> lhsWithRuleInfo =
  lhs.pattern("RuleInfo")
    .id("$ruleInfo", false)
    .constraint("businessAction == '" + businessAction + "'")
    .constraint("when == '" + place.toString() + "'")
    .end();
```

Fig. 9 The way of finding a proper business rule for evaluation in drools

changed, and now particular fact parts are represented as separate elements (subject, action, complement)—see the example below (the previously defined value attribute was preserved for backward compatibility, but it is not used). The subject represents an active element (who is taken action) while the complement is a manipulated object.

The other attribute of @BusinessRule annotation (lookup) used for identification of facts parameters remained unchanged. As one can observe, the definition of lookups is not optimized—they are repeated for every synonym. It also has to be consistent with defined glossary.

The change (introduction of the canonical form) also affected the way business rules are found for verification in Drools. Originally, the whole fact (content) was used for the rule identification. Now, each of the parts (subject, action, complement) is used separately for data filtering (see Fig. 9).

3 Case Study

Proposed extensions to the framework were tested with a case study. The testing application (see Fig. 10) was inspired by OMG SBVR standard [6] supports car renting according to the predefined business rules. It is an MVC web application which server side business logic.

The business rules were built upon the terms and facts formulated in the business vocabulary—see Fig. 11.

Business vocabulary is consistent with the source code presented by the class diagram in Fig. 12. The picture shows server-side elements (so-called backend). User interface components (web application) are not shown. Class instances are stored in a MySQL database and managed by Hibernate tool (JPA specification). *CarService* class implements some methods used for filling the database with exemplary data (e.g. *prepareCars()*, *prepareRental()*, etc.)

During implementation, a side effect of the introduced extension was observed. Vocabulary does not contain the *CarModel* term—only its role (*model*) is given. As a consequence, the connection between the class and its role is lost, so a business rule specifier has to know it exists in the source code. That problem can be easily solved by extending vocabulary grammar with the possibility of direct roles definition, e.g.

Term: model
   Description: 'Someone who rents a car'
   Role: CarModel
   Attributes: 'name', 'model'

Each business rule was separately defined in the business rule editor, and next tested with the use of test data (not presented here). If the test data met the business rule requirements, the application informs about the success; otherwise, an exception is thrown and the system informs about the rule name, which was broken.

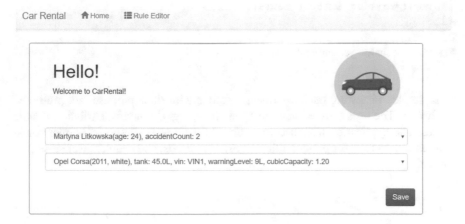

Fig. 10 Main screen of the test application

```
package  pl.edu.pwr.gawedat.carrental.model {
  Term:  customer
     Description:  "Someone  who  wants  to  rent  a  car"
     Synonyms:  'client', 'person'
  Term:  user
  Term:  order
  Term:  car
     Synonyms:  'vehicle'
     Attributes:  'tank', 'model', 'vin',
        'cubicCapacity','prodYear', 'colour'
  Term:  tank
     Synonyms:  'fuel'
  Term:  warningLevel
     Description:  'level (in litr.) of min. tank status'
  Term:  rental
     Attributes:  'customer', 'car', 'startDate', 'endDate'
  Term:  model
     Attributes:  'producerName', 'warningLevel', 'modelName'
  Term:  shop
     Attributes:  'name'
  Term:  invoice
     Synonyms:  'document'
  Term:  accidentCount
  Term:  accountBalance
  Term:  firstName
  Term:  lastName
  Term:  age

  Fact:  customer  rent  car
  Fact:  customer  has  firstName
  Fact:  customer  has  lastName
  Fact:  customer  has  accidentCount
  Fact:  customer  has  accountBalance
  Fact:  customer  has  age
  Fact:  shop  own  car
  Fact:  shop  fill  tank
  Fact:  shop  creates  invoice
  Fact:  invoice  isAbout  rental
}
```

Fig. 11 Case study—vocabulary definition

Examples of testing business rules, together with their purpose, are presented in Table 1. The tests covered new functionalities, e.g. synonyms, attributes, double values. The application response was correct for all tests. It should be mentioned that regression tests for pre-conditions, post-conditions, and enablers were also performed.

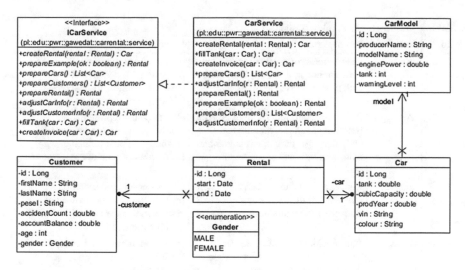

Fig. 12 Case study—model definition

4 Summary

The paper presents the architecture and the actual development state of the framework for processing behavioral business rules expressed in a controlled natural language. The framework was extended with new features. The first group of extensions is connected with the representation of business rules. It introduces syntactic sugar—shortcuts for attributes definition—as well as support for term synonyms. They required changes in the internal representation of business rules. The second group involves servicing of timers and enabling double values in conditions. The extensions were extensively tested. It is the next step to the more mature solution.

Formulation of well-defined business rules in a semi-natural language according to the specified grammar is a challenge though available support of XText tools in the editor of business rules. A user has to know business vocabulary (terms, facts), be familiar with the structure of supported rules as well as understand the limitations of the solution (e.g., the grammar of business rules could be not fully conformant with the English grammar). That problem could be at least partially addressed by the automatic generation of the vocabulary on the base of existing source code.

The proposed solution requires the source code to be properly annotated. Otherwise, business rules will not have any effect. It means that the methods in the source code, potentially connected with business rules checking, must be foreseen together with useful parameters as their navigation path is a part of the annotation definition. It is recommended to limit the number of annotations only to those most likely run to avoid empty activations of the aspect, servicing rules from the application side. Such a case could be a bottleneck and decrease application performance. To achieve this goal, collaboration with domain experts is necessary. The problems mentioned above could be partially solved by a tool checking consistency between defined vocabulary and annotations for services as well as classes representing the application model.

Table 1 Testing business rules examples

Id	Rule type	Business rule	Purpose
BR1	Pre	Rule MinimumLevelFuel is: Before person rent car, it is obligatory that tank of vehicle is at least 50;	(a) Canonical representation; (b) Synonym: person instead of customer; vehicle instead of car; (c) Attributes: tank of vehicle)
BR2	Post	Rule MoreFuelThanWarning is: After person rent vehicle, then it is obligatory that tank of car is greater than warningLevel of model of vehicle;	(a) Canonical representation; (b) Synonym: person instead of customer, vehicle instead of car; (c) Attributes: tank of car
BR3	Enabler	Rule CreateDocument is: After client rent vehicle, then it is obligatory that shop creates document using car;	(a) Canonical representation; (b) Synonyms: client instead of customer, vehicle instead of car, document instead of invoice
BR4	Timer	Rule AccidentCountGreater is: If accidentCount of customer is greater than 3, then customer cannot rent car;	(a) Timer: simplified syntax; (b) Attributes: accidentCount of customer
BR5	Timer	Rule CubicCapacityAtLeast is: If cubicCapacity of car is at least 1.6, then customer can rent car;	(a) Timer: simplified syntax; (b) Double value; (c) Attributes: cubicCapacity of car
BR6	Timer	Rule FuelAtLeast is: If tank of vehicle is at least 35.5, then customer can rent car;	(a) Timer: simplified syntax; (b) Double value (positive); (c) Attributes: tank of vehicle; (d) Synonyms: vehicle instead of car
BR7	Timer	Rule MinAccountBalance is: Only if accountBalance of customer is greater than -9.0, it is permitted that customer rent car;	(a) Timer: restricted permission syntax; (b) Double value (negative); (c) Attributes: accountBalance of customer
BR8	Timer	Rule ProdYearAtMost is: Only if prodYear of car is greater than 2008, it is permitted that a customer rent car;	(a) Timer: restricted permission syntax; (b) Integer value; (c) Attributes: prodYear of car

The rules used by the framework are written in a session object and are deleted from the Drools repository when the application is restarted. At that moment, they need to be stored manually, externally to the framework itself (however it is possible to add new rules when the application is running). In the future, the rules will be managed with the use of a database management system which will allow seeing the set of active rules and enabling/disabling selected, creation, deletion or update of existing ones.

Another problem that needs to be addressed is connected with tracing of Drools behavior. It could be solved with a logger in which information about checked rules together with their results will be stored.

References

1. Business Rule Group (2019) The business rules manifesto. http://www.businessrulesgroup. org/brmanifesto.htm. Cited 14 June 2019
2. Hnatkowska B, Kasprzyk K (2010) Business rules modularization with AOP. Przegląd Elektrotechniczny, R. 86(9), 234–238
3. Hnatkowska B, Kasprzyk K (2012) Integration of application business logic and business rules with DSL and AOP. In: Szmuc T, Szpyrka M, Zendulka J (eds) CEE-SET 2009. Springer, Berlin, pp 30–39
4. Object Constraint Language Version 2.4 (2014) OMG
5. Cemus K, Cerny T, Donahoo MJ (2015) Automated business rules transformation into a persistence layer. Procedia Comput Sci 62:312–318
6. Semantics of Business Vocabulary and Business Rules (SBVR) (2017) v. 1.4, OMG
7. Galloway J (2019) Part 6. Using data annotations for model validation. https://docs.microsoft. com/pl-pl/aspnet/mvc/overview/older-versions/mvc-music-store/mvc-music-store-part-6. Cited 14 June 2019
8. Validating Form Input. Spring by Pivotal (2019). https://spring.io/guides/gs/validating-form-input/. Cited 14 June 2019
9. Demuth B, Hussmann H, Loecher S (2001) OCL as a specification language for business rules in database applications. In: Gogolla M, Kobryn C (eds) <<UML>> 2001—The unified modeling language. Modeling languages, concepts, and tools. UML 2001. Springer, Heidelberg, pp 104–117
10. Hnatkowska B, Bień S, Ceńkar M (2012) Rapid application development with UML and Spring Roo. In: Borzemski L (at all, eds) Information system architecture and technology: web engineering and high-performance computing on complex environments. Oficyna Wydawnicza Politechniki Wrocawskiej, Wrocaw, Poland, pp 69–80
11. Bajwa IS, Lee MG (2011) transformation rules for translating business rules to OCL constraints. In: Kuester JM, Bordbar B, Paige RF (eds) Modelling foundations and applications. ECMFA 2011. Springer, Heidelberg, pp 132–143
12. Bajwa I, Bordbar B, Lee M (2010) OCL constraints generation from natural language specification. In: Proceedings—IEEE international enterprise distributed object computing workshop, EDOC. https://doi.org/10.1109/EDOC.2010.33
13. Ramzan S, Bajwa I, Haq I, Naeem MA (2014) A model transformation from NL to SBVR. In: 2014 9th international conference on digital information management, ICDIM 2014, pp 220–225. https://doi.org/10.1109/ICDIM.2014.6991430
14. NL2OCL Project (2019). http://www.cs.bham.ac.uk/~bxb/NL2OCLviaSBVR/NL2SBVR. html. Cited 20 June 2019
15. Hnatkowska B, Gawęda T (2018) Automatic processing of dynamic business rules written in a controlled natural language. In: Kosiuczenko P, Madeyski L (eds) Towards a synergistic combination of research and practice in software engineering. Studies in computational intelligence. Springer, Cham, pp 91–103
16. Framework (2019) http://www.businessdictionary.com, WebFinance, Inc. http://www. businessdictionary.com/definition/framework.html. Cited 14 June 2019
17. Bettini L (2013) Implementing domain-specific languages with Xtext and Xtend, Packt Publishing
18. The Business Rules Group (2000) Defining business rules—what they are really?, Final report, version 1.3. http://www.businessrulesgroup.org/first_paper/BRG-whatisBR_3ed.pdf. Cited 14 June 2019
19. Drools (2019) https://www.drools.org/. Cited 14 June 2019
20. Aspect Oriented Programming with Spring (2019) Pivotal software. http://docs.spring.io/ spring/docs/current/spring-framework-reference/html/aop.html. Cited 14 June 2019

Software Defect Prediction Using Bad Code Smells: A Systematic Literature Review

Paweł Piotrowski and Lech Madeyski ⓘ

Abstract The challenge of effective refactoring in the software development cycle brought forward the need to develop automated defect prediction models. Among many existing indicators of bad code, *code smells* have attracted particular interest of both the research community and practitioners in recent years. In this paper, we describe the current state-of-the-art in the field of bug prediction with the use of code smells and attempt to identify areas requiring further research. To achieve this goal, we conducted a systematic literature review of 27 research papers published between 2006 and 2019. For each paper, we (i) analysed the reported relationship between *smelliness* and *bugginess*, as well as (ii) evaluated the performance of code smell data used as a defect predictor in models developed using machine learning techniques. Our investigation confirms that code smells are both positively correlated with software defects and can positively influence the performance of fault detection models. However, not all types of smells and smell-related metrics are equally useful. *God Class*, *God Method*, *Message Chains* smells and *Smell intensity* metric stand out as particularly effective. Smells such as *Inappropriate Intimacy*, *Variable Re-assign*, *Clones*, *Middle Man* or *Speculative Generality* require further research to confirm their contribution. Metrics describing the introduction and evolution of anti-patterns in code present a promising opportunity for experimentation.

1 Introduction

Maintenance constitutes a substantial part of every software development cycle. To ensure the effectiveness of that process, a need arises to guide the maintenance and refactoring effort in a way that ensures most fault-prone components are given a

P. Piotrowski · L. Madeyski (✉)
Faculty of Computer Science and Management, Wroclaw University of Science and Technology, Wroclaw, Poland
e-mail: lech.madeyski@pwr.edu.pl

P. Piotrowski
e-mail: pawel.piotrowski@student.pwr.edu.pl

© Springer Nature Switzerland AG 2020   77
A. Poniszewska-Marańda et al. (eds.), *Data-Centric Business and Applications*,
Lecture Notes on Data Engineering and Communications Technologies 40,
https://doi.org/10.1007/978-3-030-34706-2_5

priority. Among many possible types of indicators of such "hazardous" code, code smells have attracted a lot of interest from the scientific community, as well as practitioners, in the recent years.

The notion of code smells has first been introduced by Fowler et al. [30]. They defined a set of 22 common exemplifications of bad coding practices, that can potentially signal a need for refactoring. Since then, numerous papers have been published on code smells definitions, code smell detection techniques and code smell correlation with software bugs. Some studies on software defect prediction models built with the use of machine learning techniques also attempted to evaluate the effectiveness of code smell information as a bug predictor.

A few systematic literature reviews exist regarding the topic of code smells. Freitas [31] performed a comprehensive review of the "code smell effect" to assess the usefulness of code smells as a concept. Azeem [28] reviewed machine learning techniques of code smell detection. However, an initial review of related work indicated that no exhaustive review concerning the code smells relationship with software defects and their usefulness as bug predictors have been conducted. In this paper, we attempt to perform such an investigation.

The main goals of this review are:

- to assess the state of the art in the field of fault prediction models that include code smell information,
- to evaluate the contribution of individual smell-related factors,
- to identify promising fields for further research in this area.

The rest of this paper is structured as follows. In Sect. 2 we describe related studies we were familiar with prior to conducting this review. Section 3 presents the research methodology, including research questions, search strategy and our approach to the process of study selection, quality assessment and data extraction. Section 4 presents the results of our review. In Sect. 5 we discuss and interpret these results, while in Sect. 6 we describe potential threats to the validity of our work. Section 7 concludes the review and presents potential areas for further research and experimentation.

2   Related Work

Before conducting the systematic literature review we were aware of some papers on code smell prediction and existing open access datasets of code smells. Aside from empirical studies, we also came across a few secondary studies and meta-analyses connected with the topic of code smells.

2.1   Primary Studies

Fontana et al. [27] performed an empirical comparison of 16 different machine-learning algorithms on four code smells (Feature Envy, Long Method, Data Class, Large Class) and 74 software systems, with 1986 manually validated code smell samples. They found that the highest performances were obtained by J48 and Random Forest, while detection of code smells can provide high accuracy (over 96%).

Palomba et al. [37] contributed the data set containing 243 instances of five smell types from 20 open source projects manually verified by two MSc students. They also presented LANDFILL, a web-based platform aimed at promoting the collection and sharing of code smell data sets.

Palomba et al. [36] used the data set presented in [37] and proposed Historical Information for Smell deTection (HIST) approach exploiting change history information to detect instances of five different code smells (Divergent Change, Shotgun Surgery, Parallel Inheritance, Blob, and Feature Envy). The results indicate that the precision of HIST is between 72 and 86%, while its recall is between 58 and 100%.

2.2 Secondary Studies

Although a number of systematic literature reviews related with the topic of code smells exist, we found only two secondary studies reporting on their influence on bug prediction models, of which only [29] is a systematic literature review.

Cairo et al. [29] analysed 16 empirical studies to examine to what extent code smell detection influences the accuracy of bug prediction models. Their study focused primarily on the types of code smells used in the experiments, as well as the tools, techniques and resources used by researchers to find evidence of the influence of code smells on the process of fault prediction.

Gradišnik and Heričko [32] performed a review of 6 research papers published in the period from 2006 to 2014 to determine whether any subgroups of code smells influence the quality of software in a particularly harmful way. They studied the correlation between 22 distinct types of code smells and the fault-proneness of classes. Their analysis indicated weak and sometimes contradictory correlations between individual smells and the bugginess of classes. Also, they found that researchers often focus on subgroups of smells rather than analyse their full range, their choices usually being arbitrary.

An extensive systematic review regarding "the code smell effect" has been performed by Freitas et al. [31]. In a survey of 64 primary studies, the researchers synthesised how the concept of code smells influences the software development process. In their results, they indicated some inconsistencies between the findings of different studies. While analysing the role of people in smell detection, they found that evaluation of smells by humans has many flaws. They also created a thematic map concerning code smell themes such as "correlation with issues of development", "human aspects", "programming" and "detection".

The only available systematic review on the subject written by Cairo et al. [29] was the starting point in our research. We found that an extension of the analysis is required, as the set of studies covered in the review did not include some relevant papers we found manually prior to this review (such as [17] by Palomba et al. and their studies concerning code smell intensity). The review also did not provide a synthesis enabling us to survey the already investigated smells and extract the under-researched bug prediction factors. For that reason, we decided an extended systematic literature review on the topic of the influence of code smells on bug prediction is required.

3   Research Methodology

In this section we present the methodology of our systematic literature review. As suggested by Kitchenham et al. [33], this includes research questions driving the review process, description of the search strategy, selection process, as well as the approach to quality assessment and data extraction. It is important to note that throughout this paper we use the words *faults*, *defects* and *bugs* interchangeably.

3.1   Research Questions

Our systematic literature review aims to answer the following research questions.

RQ1—How does code smell detection influence the accuracy of defect-prediction?
RQ2—Which metrics and code smells are most useful when predicting defects?

3.2   Search Strategy

Our initial set of primary studies consisted of four research papers [17, 22, 34, 35]. These papers have been found in a manual search for literature on code smells. Next, the set of primary papers from two literature reviews [29, 32] was considered. These two reviews included 19 distinct papers (16 in [29] and 6 in [32] with 3 titles overlapping), but none of the four papers from our manual search. Knowing that the four papers are relevant to our research, we decided an extended primary studies search and a broader literature review must be conducted to appropriately describe the current state-of-the-art.

To extend the literature search, we decided on two additional search methods. First, we performed an automated search in the IEEE Xplore Digital Library.[1] After selecting the relevant results, we also performed forward and backward snowballing on these articles.

A search query using boolean operators was obtained for the automated search procedure. Its major expressions have been derived directly from our research questions. It consisted of a concatenation of four major terms using the operator **AND**, with possible different spellings and synonyms of each term concatenated with the use of the operator **OR**.

---

[1] https://ieeexplore.ieee.org/Xplore/home.jsp.

Table 1 Search results and data sources

Source	Papers found	Relevant	Finally selected
Manual search[2]	30	16	16
IEEE Xplore digital library	47	10	9
Snowballing	3	3	2
Total	80	29	27

[2] Includes papers covered by [29, 32]

The search string has been defined as follows:

(*software* OR "*software project*" OR "*software projects*")

AND

("*code smell*" OR "*bad smell*" OR "*code smells*" OR "*bad smells*" OR *antipattern* OR *antipatterns* OR *anti-pattern* OR *anti-patterns* OR "*bad design*" OR "*design flaw*")

AND

(*bug* OR *bugs* OR *fault* OR *issue* OR *failure* OR *error* OR *flaw* OR *defect* OR *defects*)

AND

(*predicting* OR *prediction* OR "*prediction model*" OR *identification*)

The results of the search are presented in Table 1. The detailed depiction of the search and selection process is presented on Fig. 1.

3.3   Selection Process

First, all the results retrieved from the automated search have been preliminarily reviewed based on their titles and abstracts. While deciding on the inclusion or rejection of each article, the following criteria have been taken into consideration.

Inclusion criteria:

- Articles describing the correlation between code smells and software defects
- Articles reporting on fault prediction models based on code smell detection
- Articles which examine the influence of code smell data used as an additional predictor in other software fault prediction models
- Articles mentioning improvements to existing smell-based bug prediction techniques.

Exclusion criteria:

- Articles not written in English
- Articles not related to software engineering

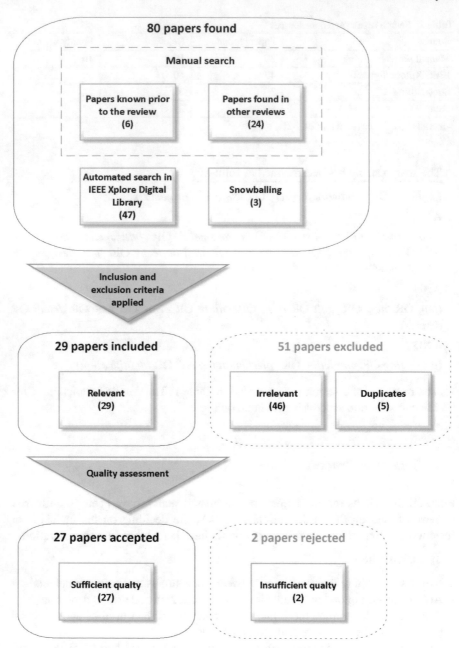

Fig. 1 The search and selection process

Table 2 Quality assessment criteria

Q1 Is the paper based on empirical evidence?
Q2 Is the research objective clearly stated?
Q3 Are the used code smells clearly defined?
Q4 Are the names of analysed software projects specified?
Q5 Does the paper state how the metrics and/or code smells data was collected?
Q6 Does the paper state the source of fault-proneness data?
Q7 Does the paper evaluate the predictive power of considered code smells in a manner allowing to draw conclusions about their influence on fault-proneness?
Q8 Is the experiment presented in the paper reproducible?
Q9 Is the conclusion clearly stated?

- Articles that focus on code smells detection techniques
- Articles focusing on the human aspect of introducing code smells.

On each accepted article forward and backward snowballing has been performed. The same inclusion/exclusion criteria have been applied on the results, which lead to including 3 additional articles [9, 12, 38].

3.4 Quality Assessment

To assess the quality of the selected studies, we developed a checklist of quality criteria, presented in Table 2. Its initial version was based on other literature reviews, but the checklist evolved during the initial stage of the assessment.

For each question, three possible grades were assigned—1.0(Yes), 0.5(Partially) or 0.0(No). The grades were inserted into a spreadsheet, where they were then summed up to produce a numerical indication of each paper's quality and 'suitability" for our review.

After performing the assessment it was established that only two of the considered papers are not suitable for the review, as they contained an insufficient amount of information. These two papers had the lowest quality score of 5.0 and 5.5. This way, a minimal threshold for the paper quality was established at the value 6.

3.5 Data Extraction

Once the final selection of the papers was performed, we proceeded onto extracting data relevant for the review. This included reference attributes (such as title, authors, no of pages), as well as data necessary to answer our research questions. The data extraction form attributes are presented in Table 3.

Table 3 Data extraction form

i	Title
ii	Names of the authors
iii	Year of publication
iv	Number of pages
v	Research questions
vi	Answers to research questions
vii	Types of code smells used
viii	Other metrics used
ix	Programming language
x	Analysed software projects
xi	Source of metric/code smell data
xii	Fault-proneness evaluation method
xiii	Conclusion about the influence of code smells on bug predictionc
xiv	Limitations

While analysing each paper's research questions and their answers, we were primarily interested in the researchers' conclusions about the influence of code smell detection on bug prediction. In the field "Conclusion about the influence of code smells on bug prediction" we summarised this judgement to indicate whether the paper found a connection between the two or not.

We were also interested in the types of code smells used in each study, the source of data concerning faults and the statistical methods used to check whether a connection between the two exists.

4 Results

In this section, we present the results of our review. First, we briefly describe the demographics of the chosen papers. After that, we describe the results with respect to each of the research questions.

4.1 Demographics

The full list of papers selected for the review is presented at the end of this article. Most of the research papers in this area have been published in conferences. The earliest analysed paper is from 2006. The vast majority of the studies (22 out of 27) were published after 2011, which corresponds to the software engineering community's rising interest into the impact of code smells on software projects.

4.2 Code Smells and Bugginess

In this section we present the general conclusions of the reviewed papers concerning the influence of code smells on defect-proneness, irrespectively of the code smell type. The individual kinds of smells examined in the analysed studies are described separately in Sect. 4.3.

No paper provided conclusive evidence indicating that code smells are the direct source of faults. However, all of the examined papers studied one or both of the following dependencies:

- the statistical correlation between code smells and software bugs
- the influence of smell detection on the performance of bug prediction models.

A. Correlation between the presence of code smells and faultiness

In general, studies aiming at exploring the correlation between code smells and bugs performed different statistical tests to check whether classes, methods and modules containing code smells contained also more defects. Table 4 presents the implications of their findings.

Table 4 Correlation between smells and bugs

Paper ID	Positive	No link	Negative
[26]	X		
[21]	X		
[35]	X		
[8]	X		
[23]	X		
[1]	X		
[19]		X	
[4]	X		
[6]		X	
[2]	X		
[10]	X		
[16]	X		
[15]	X		X
[5]	X	X	
[25]		X	
[20]	X		
[12]	X		
[9]	X		
[14]	X		
[7]	X		
[13]	X		

Zhang et al. [26] studied the influence of 13 types of code smells on the faultiness of 18 versions of Apache Commons series software. They analysed each code smell separately and concluded that there is a positive relationship between bad design and defect-proneness. However, they pointed out that some smells point to software defects more than others.

The authors of [21] showed in a brief study that a strong, significant relationship exists between 3 types of smells and class errors of Eclipse 2.1.

Ma et al. [35] found slight statistical agreements (Cohen's Kappa statistic in range [0.01..0.20]) between the results of fault detection and the detection of 3 out of 8 analysed code smells.

Jaafar et al. [8] examined whether classes having static relationships with smelly classes are more defect prone. The authors conducted their studies using 11 types of code smells and found that such relationships often indicate a higher level of bug-proneness.

In their study on the influence of code smell detection on bug prediction models, the authors of [23] performed an analysis of the co-occurrence of bugs and faults in the source code of 21 versions of two Java software projects. They found that the density of bugs (that is the number of bugs normalized with respect to file size) is generally higher in files with antipatterns as compared to files without them. They performed a Wilcoxon rank sum test, which provided statistically significant results concerning this correlation for 17 out of the 21 software versions analysed.

The study conducted in [1] focused on comments and whether densely commented modules contain more faults than uncommented ones. The study analysed 3 software projects (Eclipse, Apache Tomcat, Apache Ant) and concluded that there is in fact a positive correlation between the number of comments and the number of faults. This study was extended in [2], where the authors confirmed that well-commented modules were on average 1.8 to 3 times more likely to contain bugs. However, the study results indicated that more comments do not necessarily mean more faults.

In [19], the authors explored whether the *Clones* code smell indicates a higher level of defect-proneness. In their study, they analysed monthly snapshots of repositories of 4 software products written in C and found no such relationship. In fact, they discovered that in many cases cloned code contains on average less defects and introducing more clones into the code does not make it more buggy. They concluded that clones should not be regarded as a smell, as they are not detrimental to software quality.

In [4], the authors attempted to determine the association between anti-patterns, maintainability and bugs. In their analysis of 34 Java projects, they found a positive correlation between the the number of faults and the number of anti-patterns (with Spearman correlation 0.55 with p below 0.001).

Hall et al. [6] took on five less popular code smells and investigated each one's influence on defects. They obtained mixed results and no definite positive or negative correlation was found. The authors indicated that a major threat in this research area is posed by inconsistencies in the definitions of code smells between the studies.

Jaafar et al. [10] explored the fault-proneness of classes where anti-patterns co-occur with clones. "Co-occurring" in this context did not mean that the cloned code necessarily included anti-patterns. It meant that such a class contained both some anti-patterns and some cloned code. Results of the performed study indicated that classes with such co-occurrences were at least 3 times more likely to be faulty in comparison with other classes.

Palomba et al. [16] reported a study on the diffuseness of code smells and their influence on change- and bug-proneness based on an analysis of 30 software projects. The authors concluded that code smells have a big effect on change-proneness and a medium effect on defect-proneness. Moreover, a higher number of code smells directly indicates a higher level of fault-proneness. The paper also observed that code smells are not necessarily a direct cause of faults, but a "co-occurring phenomenon".

In [15], the authors investigated two code smells related to size—*God Class* and *Brain Class*. They found that classes containing these two smells had more defects than those without them. However, when normalized with respect to size, these classes had less defects "per line of code". Hence, the overall conclusion about the correlation of these smells with defects is not straightforward.

In [5], the authors reported a positive correlation between occurrences of 5 types of code smells and software defects. However, no design flaw correlated with defects more than others. The levels of correlation varied widely, depending on the analysed software project.

A multiple case study was conducted in [25], where researchers registered the development process of four different, unspecified Java projects. The projects have been developed by six developers for a period up to four weeks. Code smells were registered before the maintenance phase and an investigation was performed to establish whether the observed faults could be caused by the smells. The authors found no causal relationship, as only around 30% of faulty files contained smells. The study also determined that introducing new smells did not increase the proportion of defects.

Saboury et al. [20] explored the influence of 12 code smells on the "survival time" of JavaScript files. Survival time was defined as the average time until the first occurrence of a bug. The paper reported that hazard rates of the files without code smells, calculated based on their survival times, were on average 65% lower than those with smells.

In [12], the authors explored the relationship between code smells and change- and defect-proneness, as well as whether anti-patterns are in any way related to size. In their analysis of four open source Java software projects they found that the existence of anti-patterns does indeed indicate a higher level of change and defect-proneness. However, they noted that class size itself does not influence bugginess. Their study was extended in [9], where the process of anti-pattern mutation was analysed. By mutation the authors meant the evolution of anti-patterns over time, especially evolution of an antipattern into a different type of antipattern. The results of this analysis indicated that anti-patterns that do not mutate are more fault-prone that the ones that are structurally altered over time.

Marinescu and Marinescu [14] investigated the relationship between class-clients of smelly and non-smelly classes. Four types of code smells were considered. The researchers found that classes that use smelly classes are on average more defect-prone. Although such a positive correlation was found, no direct causation was proved.

In [7], the authors explored four different techniques of technical debt detection. Apart from code smells, the other analysed techniques were modularity violations, grime buildup and automatic static analysis. The paper set out to find whether these techniques result in the same classes being classified as problematic, as well as examine the correlation between their findings and fault-proneness. The study concluded that not much overlap between all the techniques was found, but of the 9 types of code smells studied, 2 of them proved to be well-correlated with buggy classes.

In [13], the authors investigated in a brief study the relationship between occurrences of 6 types of code smells and the error proneness of 3 versions of Eclipse. They found that while the influence widely varies between different smell types, smelliness in general does point to fault-proneness.

In summary, out of 21 papers examining the relationship between the presence of code smells and the presence of bugs, 16 reported on the existence of a positive correlation between the two. Three papers found no such link, while two papers reported inconclusive or conflicting results.

B. Influence of code smell detection on bug prediction performance

Some of the studies investigated how the information about the smelliness of classes and methods influences the accuracy of bug prediction models. Table 5 shows the general conclusions made by each study concerning their effect.

Ubayawardana and Damith Karunaratna [24] used a bug prediction model based on 17 source code metrics as a base model. The authors introduced 4 additional, anti-pattern related metrics (relating to amount, complexity and recurrence of anti-patterns), as well as an intensity index. They compared the performance of the model with and without the smell-related metrics using three different classifiers trained on 13 different software projects. Their results showed that bug prediction based solely on source code metrics is not reliable. However, including the anti-pattern metrics significantly enhanced the performance of the models. In some cases, with the use of the Random Forest classifier, the authors reported achieving 100% accuracy of bug prediction on test data.

Table 5 Influence on bug prediction accuracy

Paper ID	Positive	Neutral	Negative
[24]	X		
[35]	X		
[11]		X	
[23]	X		
[17]	X		
[18]	X		
[4]	X		
[3]	X		
[22]	X		

The study conducted in [35] found only a small agreement between code smell detection and the results of fault prediction using a multivariable logistic regression model based on cohesion metrics. However, the authors noted that using code smell detection results can improve the recall of bug prediction, as smelliness often points to faultiness. Using combined data of all the considered code smells, the fault prediction results were improved by 9 to 16%.

In [11], the authors performed an analysis of how sampling and resampling techniques influence the performance of bug prediction models based on source code metrics and code smells. Their study concluded that code smells themselves are not a better predictor than metrics.

Taba et al. [23] used as a base model a logistic regression bug prediction model based on source code metrics that proved to be accurate in previous studies. They developed four smell-related metrics of their own and added them to the model, one by one, to test its performance. The best results were achieved with the use of their "Anti-pattern Recurrence" metric, which proved to improve the accuracy of the baseline model by 12.5% on average.

In [17], the authors introduced a code smell intensity index as a bug predictor. For their baseline model they used 20 code metrics proposed by other researchers. Their study covered how the introduction of smell intensity affects the performance of the base model. They found that code smell intensity always positively influences the accuracy of bug prediction. Furthermore, their analysis indicated that smell intensity provides additional information that a simple indicator of smell presence could not give. The authors continued this study in [18], where they verified the influence of their intensity index against models based on product and process metrics, as well as metrics related to the history of code smells in files. Their results indicated that smell intensity is an important predictor and its use always increases the accuracy of bug prediction.

In [4], while analysing the impact of code smells on software maintainability, the authors observed that smells also show significant predictive power. They noted that the number of anti-patterns appearing in a class is only a slightly worse predictor than the intricate metric model they have studied as their main subject.

Aman et al. [3] investigated whether *Lines Of Comments* (LCM) can serve as a useful metric for predicting defects in software. They used this metric to categorize methods as more-commented or less-commented. They found that more-commented methods are on average 1.6–2.8 as likely to contain bugs. Furthermore, they used existing fault data to train logistic regression models and found that LCM is a useful bug predictor, alongside metrics related to the method's size and complexity.

In [22], the authors built defect prediction models with the use of code smell and churn metrics. They investigated and compared the impact of those metrics on bug prediction performance. They reported that code smell metrics significantly improved the performance of the models. The machine learning algorithms they used, Logistic Regression and Naive Bayes, also performed better with code smell metrics than churn metrics.

To sum up, eight out of nine studies covering the effect of code smell-related information on the accuracy of bug prediction reported positive results. The one remaining study described the code smells' defect prediction ability as no greater than the one of source code metrics.

4.3   The Impact of Individual Smells

A. Landscape of studied code smells

Table 6 presents smells and smell-related metrics studied in the analysed research papers. The smells are sorted in a descending order starting from most frequently studied.

The most popular smells are those connected with large size and high complexity of classes and methods, such as *Blob*, *God Method*, *Complex Class* or *Long Parameter List*. Other frequently studied smells are related to improper inheritance (*Refused Parent Bequest*), improper handling of class attributes (*Class Data Should Be Private*) and bad encapsulation (*Feature Envy*, *Lazy Class*, *Shotgun Surgery*).

Among the less studied anti-patterns there are such smells as *Inappropriate Intimacy*, *Base Class Should Be Abstract*, *Temporary Field*, *Middle Man*, *Switch Statements* or *Data Clumps*. Also, the set of anti-patterns used in [20] contains many unique smells, as the authors used some JavaScript-specific smells, taken from popular JavaScript coding guides.

B. Usefulness and popularity of smells

While some papers treated code smells collectively, using as a predictor an indication of whether any type of smell was present in a given class or method, most studies evaluated them separately. Studies of the second type were often able to draw conclusions about which types of smells proved to be most useful in the task of bug prediction. Table 7 summarises those conclusions, including cases where researchers found little or close to no influence of smells on the level of software faultiness.

The information from Tables 6 and 7 has been aggregated into Table 8. This table combines information concerning all the code smells and metrics used as bug predictors in the analysed studies and allows to evaluate their popularity as well as potential usefulness.

The code smell that was most frequently indicated as highly correlated with bugs was *God Class*. Researchers reported high usefulness of this smell in 8 out of 18 analysed studies. Together with high reported usefulness of *God Method* and *Brain Class* this indicates that big sizes and high complexities of classes and methods are both popular and effective in the task of bug prediction. One notable exception from this trend was presented in [15]. In this study, authors found that *God Classes* and *God Methods* do have on average more faults than other classes. However, they also found that when the number of bugs is normalized with respect to the size of classes or methods, they are in fact less buggy. Their study pointed out that most studies analysing *God Class* and *God Method* take into consideration absolute, not relative values of bug-proneness of those classes.

Table 6 Smells studied in the analysed papers

Smell (smell metric)	Paper IDs
Blob/God class	[4, 7–18, 21, 23, 25, 26, 35]
Long/God/Brain method	[4, 5, 8–13, 15–18, 20, 21, 23, 25, 26]
Brain/Complex class	[7–10, 12, 14–16, 20, 23, 26, 35]
Refused parent bequest	[4, 7–10, 12, 13, 16, 23, 25, 26, 35]
Long parameter list	[4, 8–10, 12, 16, 20, 23, 26, 35]
Message chain	[6, 8–10, 12, 16–18, 23, 26]
Class data should be private	[8–10, 12, 16, 23, 26, 35]
Feature envy	[4, 5, 7, 11, 13, 14, 16, 25]
Lazy class	[4, 9, 10, 12, 16, 23, 26, 35]
Shotgun surgery	[4, 5, 7, 13, 17, 18, 21, 25]
Speculative generality	[6, 8–10, 12, 16, 23, 26]
Anti singleton	[8–10, 12, 23, 26, 35]
Data class	[7, 10, 13, 14, 17, 18, 25]
Swiss army knife	[8–10, 12, 23, 26, 35]
Large class	[4, 9, 10, 12, 23, 26]
Spaghetti code	[8, 10, 12, 16, 23, 26]
Dispersed coupling	[5, 7, 17, 18]
Clones/Duplicated code	[10, 19, 25]
Comments (LOC)	[1–3]
Data clumps	[6, 25]
Intensive coupling	[5, 7]
Middle man	[6, 16]
Switch statement	[6, 20]
Variable Re-assign	[20, 25]
Base class should be abstract	[10]
Empty catch block, dummy handler, unprotected main programs, nested try statement, careless cleanup, exceptions thrown from finally block	[11]
Inappropriate Intimacy	[16]
Lengthy lines, Chained methods, Nested callbacks, Assignment in conditional statements, Extra bind, This assign, Depth	[20]
Misplace class	[25]
Temporary field	[4]
Tradition breaker	[7]
unknown	[22]
Use interface instead of implementation, Interface segregation principle violation	[25]

Table 7 Papers reporting on smell usefulness

factor	Papers reporting high usefulness	Papers reporting low usefulness
Blob/God class	[7, 8, 13, 16–18, 26, 35]	[15]
Long/God/Brain method	[12, 17, 18, 21, 26]	
Message chain	[9, 12, 17, 18]	
Dispersed coupling	[7, 17, 18]	
Comments (LCM)	[1–3]	
Brain/Complex class	[8, 12, 26]	[15]
Anit-pattern/smell intensity	[17, 18, 24]	
Avg no of anti-patterns	[4, 18, 24]	
Refused parent bequest	[9, 16, 35]	[13, 21]
Shotgun surgery	[18, 21]	
Anti-pat. recurrence length	[23, 24]	
LinesOfCode (size)	[3, 26]	[6]
Data class	[17, 18]	[13, 21]
Anti singleton	[12]	
Assignment in cond. statements	[20]	
Clones/Duplicated code	[10]	
Inappropriate intimacy	[16]	
Large class	[21]	
Long Parameter list	[26]	
Variable Re-assign	[20]	
CallDependency	[26]	
Cyclomatic complexity	[3]	
Anti-pat. complexity	[24]	
Anti-pat. cumulative pairwise diff.	[24]	
Anti-pat. indicator (present/absent)	[10]	
Lazy class	[12]	[35]
Swiss army knife	[8]	[35]
Feature envy	[16]	[13, 21]
CDSBP		
Spaghetti code		
Middle man		[6]
Speculative generality		[6]
Switch statement		[6]

Table 8 Number of papers reporting on individual smells—summary Legend: **s**: code smell **m**: metric; **#+**: number of papers reporting high usefulness **# –**: number of papers reporting low usefulness **#papers**: total number of papers reporting

Type	Factor	# +	# –	# Papers
s	Blob/God class	8	1	18
s	Long/God/Brain method	5	0	17
s	Message chain	4	0	10
s	Brain/Complex class	3	1	12
s	Dispersed coupling	3	0	4
s	Comments (LCM)	3	0	3
m	Anti-pattern/smell intensity	3	0	3
m	Avg no of anti-patterns	3	0	3
s	Refused parent bequest	3	2	12
s	Shotgun surgery	2	0	8
m	Anti-pattern recurrence length	2	0	2
m	LinesOfCode (size)	2	1	3
s	Data class	2	2	7
s	Long Parameter list	1	0	10
s	Anti singleton	1	0	7
s	Large class	1	0	6
s	Clones/Duplicated code	1	0	3
s	Variable Re-assign	1	0	2
m	Anti-pattern complexity	1	0	2
m	Anti-patten indicator (present/absent)	1	0	2
s	Assignment in conditional statements	1	0	1
s	Inappropriate intimacy	1	0	1
m	Call dependency	1	0	1
m	Cyclomatic complexity	1	0	1
m	Anti-pattern cumulative pairwise differences	1	0	1
s	Lazy class	1	1	8
s	Swiss army knife	1	1	7
s	Feature envy	1	2	8
s	CDSBP	0	0	8
s	Spaghetti code	0	0	6
s	Speculative generality	0	1	8
s	Middle man	0	1	2
s	Switch statement	0	1	2

Other popular smells concerned the inter-relations of classes and the communication between them. *Message Chains* was concerned a good indicator of faultiness in 4 papers (out of 10 covering this anti-pattern). *Dispersed Coupling* and *Refused Parent Bequest* were both described as useful by 3 out of 12 papers that concerned these smells. However, Shatnawi and Li [21] regarded *Refused Parent Bequest* as too infrequent to analyse, while Li and Shatnawi [13] concluded that its influence on error rates is very small.

Aman et al. [2] analysed the influence of *Comments* in studies [1–3]. All those studies indicated a high correlation between well-commented and buggy modules (1.6-3 times higher than in other classes). The authors point out that this makes *Comments* a useful indicator of bugginess.

Extensive studies conducted by Palomba et al. [17] in [17, 18] proved that code *smell intensity* is a useful bug predictor. The inclusion of *smell intensity* as a predictor improved the performance of bug prediction models based on process, product and anti-pattern metrics. The usefulness of *smell intensity* as a predictor was also confirmed by the empirical study performed in [24].

A similar metric, *Average number of anti-patterns* was studied in [4, 18, 24]. In all three studies the authors concluded that information about how smelly a class is, in the sense of how many code smells it contains, is also useful in bug prediction.

Some popular smells proved to have a much smaller effect on bug-proneness. Smells such as *Long Parameter List*, *Anti-singleton* and *Large Class* were deemed useful only by one paper, despite being covered subsequently by 10, 7 and 6 papers each.

The *Data Class* smell was considered a very useful predictor by [17, 18]. However, Shatnawi and Li [13, 21] did not find a significant correlation between this smell and bug-proneness.

In their research, Hall et al. [6] showed that the *Switch Statement* smell has no effect on bug-proneness of any of the studied software projects. *Middle Man* and *Speculative Generality* provided mixed results that varied depending on the project.

Conflicting results were reported regarding smells *Lazy Class*, *Swiss Army Knife* and *Feature Envy*. Despite these smells being studied relatively often, only singular papers were able to reach any clear (although often contradictory) conclusions about their predictive power. Other quite popular smells, such as *Class Data Should Be Private* and *Spaghetti Code* (described in 6 and 8 articles respectively) were not singled-out even once, their influence usually being present but minimal.

5 Discussion

In this section we summarise the findings of our review and address the proposed research questions.

RQ1. How does code smell detection influence the accuracy of bug-prediction?

The great majority of analysed research papers found a positive correlation between code smells and software bugs. Although the extent of how strong and significant this relationship is varied depending on the smell types and software projects analysed, a general conclusion can be drawn that code smells are a potentially good predictor of software defects. It needs to be noted, however, that a substantial number of studies found no direct link between the studied types of smells and the bugginess of analysed software.

Many studies attributed these varying levels of code smell harmfulness in different projects to distinct company policies and software engineering practices employed in the process of software development. With different development practices, the same code smells may induce detrimental effects on software quality or have no influence at all.

Another cause of singular contradictory results might be the large diversity of code smell definitions, as well as different fault and smell data gathering methods/tools. Lack of unified methodology makes comparing the results of different studies difficult.

Synthesising the results of studies focusing on the predictive power of code smells in bug prediction models allowed to reach a clearer conclusion. Nearly all analysed papers concluded that the inclusion of code smell information improves the accuracy of bug prediction in virtually any previously devised prediction model. The smallest effect was reported by a study that described the influence of code smells as comparable to that of source code metrics. This fact reaffirms the usefulness of code smells as bug predictors.

RQ2. Which metrics and code smells are most useful when predicting defects?

Code smells relating to extensively big classes and methods of high complexity, such as *God Class* and *God Method* were not only the most frequently studied, but were also the smells that proved to be very well correlated with software bugs in the largest number of empirical studies. The interest into those smells, as well as reports of their effectiveness correspond with the common belief of the software development community that writing large code modules of high complexity is a bad practice.

Message Chains and *Dispersed Coupling* smells were relatively often singled out as exceptionally good bug predictors. Out of 9 studies reporting a positive influence of a set of code smells including *Message Chains*, 5 of them underlined this anti-pattern's exceptionally high positive correlation with bugs. In the case of *Dispersed Coupling*, this ratio was 3 out of 4.

Also *Comments* proved to be valuable bug predictors, reaffirming the common clean code practice that source code should be self-explanatory. Although the number of comments itself was not directly related to the number of bugs, the in-module comment presence indicator proved to be a valuable predictor, signalling pieces of code that are overly complicated, unclear and potentially faulty.

Good bug prediction results were also achieved by anti-pattern metrics. Especially *code smell intensity* deserves to be mentioned, as the extensive studies investigating its bug prediction power reported it to increase the accuracy of all common types of previously tested bug prediction models, reaching exceptionally good results.

Also *Average number of anti-patterns* and *Anti-pattern recurrence length* smell-related metrics proved to be successfully used as bug predictors alongside other factors.

6 Threats to Validity

The primary threat to validity in any systematic literature review is related to the completeness of the set of analysed studies. We put particular attention to developing an elaborate search term in order to obtain a comprehensive set of potentially relevant papers. However, the set of analysed papers could be biased due to limiting our automated search to the IEEE Xplore Digital Library. We chose this database based on our experience with the results of a few pilot searches. IEEE results included all the papers we were previously familiar with, as well as relevant papers we found in other databases such as ACM and ProQuest. IEEE also performed well with long search terms and consistently provided concise and relevant results. Any singular studies related with our research questions that we could have left out due to the above limitations should not invalidate the overall conclusions of our review, taking into consideration its qualitative nature.

Another factor that could influence the validity of this review is positive publication bias. Of all the relevant papers we have found, only a small minority reports negative results which undermine the usefulness of code smells as bug predictors. In our review we assumed that this trend corresponds with the reality and the majority of studies reporting positive results can be treated as an indicator of an existing correlation between smells and bugs.

It is also worth to note a few recurring validity threats of the analysed papers themselves. Although they are not related to the methodology of our review, they can be in some way related to its conclusions. Most papers dealt only with Java projects. Single studies analysing C, JavaScript or PHP projects took into consideration code smell types that differed from the ones used in Java. An insufficient amount of research has been conducted so far to determine whether the smells-bugs dependency is similar in projects written in other programming languages and draw conclusions about its generality.

Also, many studies did not conduct any tests to verify the accuracy of smell detection. The authors generally leaned on other research papers proving the accuracy of chosen smell detection tools or devised their own, metric-based smell finders. All the analysed studies assumed that the results obtained by their smell detection techniques were accurate.

7 Conclusions

This systematic literature review aimed at describing the state-of-the-art of the research on the use of code smells in bug prediction, as well as providing an analysis of the usefulness of different code smell-related factors. To achieve this aim, we conducted an analysis of 27 research papers, paying particular attention to the information concerning:

- the general contribution of smell detection to bug prediction models,
- the variety of used smells and metrics and their usefulness,
- the areas of related research that remain insufficiently explored.

The results of our work show that code smells are indeed a good indicator of bugs. However, their usefulness differs depending on the type of anti-pattern and metric analysed, as well as the software project in which they are tested. Among all the analysed types of smells, *God Class*, *God Method* and *Message Chains* stand out. Very good results were also obtained with the use of smell-related metrics, especially *Code smell intensity*.

Our survey also disclosed a large group of existing code smell types that remain very scarcely researched. Smells such as *Inappropriate Intimacy*, *Variable Re-assign* and *Clones* showed promising bug-predicting properties in individual studies, but this effect requires further validation.

Middle Man and *Speculative Generality* were only analysed in two empirical studies, providing inconclusive results. Their usefulness in bug prediction is also a potential field for further research.

Good results were achieved with the use of metrics describing the structure and intensity of code smells found in software systems, as well as the history of their introduction and evolution. This type of data, along with information regarding the coupling and co-occurences of different types of code smells might constitute a good field for experiments and can provide an opportunity to develop a new, useful code smell-related metric to use as a bug predictor.

Papers analysed in the systematic literature review

1. Aman H (2012) An empirical analysis on fault-proneness of well-commented modules. In: Proceedings - 2012 4th international workshop on empirical software engineering in practice, IWESEP 2012, pp 3–9. https://doi.org/10.1109/IWESEP.2012.12
2. Aman H, Amasaki S, Sasaki T, Kawahara M (2014) Empirical analysis of fault-proneness in methods by focusing on their comment lines. In: Proceedings—Asia-pacific software engineering conference, APSEC. https://doi.org/10.1109/APSEC.2014.93
3. Aman H, Amasaki S, Sasaki T, Kawahara M (2015) Lines of comments as a noteworthy metric for analyzing fault-proneness in methods. IEICE Trans Inf Syst. https://doi.org/10.1587/transinf.2015EDP7107
4. Bán D, Ferenc R (2014) Recognizing antipatterns and analyzing their effects on software maintainability, vol 8583 LNCS. https://doi.org/10.1007/978-3-319-09156 3_25

5. D'Ambros M, Bacchelli A, Lanza M (2010) On the impact of design flaws on software defects. In: Proceedings - international conference on quality software 1:23–31. https://doi.org/10. 1109/QSIC.2010.58
6. Hall T, Zhang M, Bowes D, Sun Y (2014) Some code smells have a significant but small effect on faults. ACM Trans Softw Eng Methodol. https://doi.org/10.1145/2629648
7. Izurieta C, Seaman C, Cai Y, Shull F, Zazworka N, Wong S, Vetro' A (2013) Comparing four approaches for technical debt identification. Softw Qual J 22(3):403–426. https://doi.org/10. 1007/s11219-013-9200-8
8. Jaafar F, Guéhéneuc Y, Hamel S, Khomh F (2013) Mining the relationship between anti-patterns dependencies and fault-proneness. In: 2013 20th working conference on reverse engineering (WCRE), pp 351–360. https://doi.org/10.1109/WCRE.2013.6671310
9. Jaafar F, Khomh F, Gueheneuc YG, Zulkernine M (2014) Anti-pattern mutations and fault-proneness. In: Proceedings - international conference on quality software, pp 246–255. https:// doi.org/10.1109/QSIC.2014.45
10. Jaafar F, Lozano A, Gueheneuc YG, Mens K (2017) On the analysis of co-occurrence of anti-patterns and clones. In: Proceedings - 2017 ieee international conference on software quality, reliability and security, QRS 2017. https://doi.org/10.1109/QRS.2017.38
11. Kaur K, Kaur P (2017) Evaluation of sampling techniques in software fault prediction using metrics and code smells. In: 2017 international conference on advances in computing, communications and informatics, ICACCI 2017, vol 2017-Janua. IEEE, pp 1377–1386. https://doi. org/10.1109/ICACCI.2017.8126033
12. Khomh F, Penta MD, Guéhéneuc YG, Antoniol G (2012) An exploratory study of the impact of antipatterns on class change- and fault-proneness. Empir Softw Eng 17(3):243–275. https:// doi.org/10.1007/s10664-011-9171-y
13. Li W, Shatnawi R (2007) An empirical study of the bad smells and class error probability in the post-release object-oriented system evolution. J Syst Softw 80(7):1120–1128. https://doi. org/10.1016/j.jss.2006.10.018
14. Marinescu R, Marinescu C (2011) Are the clients of flawed classes (also) defect prone? In: Proceedings - 11th IEEE international working conference on source code analysis and manipulation, SCAM 2011, pp 65–74. https://doi.org/10.1109/SCAM.2011.9
15. Olbrich SM, Cruzes DS, Sjøberg DI (2010) Are all code smells harmful? a study of god classes and brain classes in the evolution of three open source systems. In: IEEE international conference on software maintenance. ICSM. https://doi.org/10.1109/ICSM.2010.5609564
16. Palomba F, Bavota G, Penta MD, Fasano F, Oliveto R, Lucia AD (2018) On the diffuseness and the impact on maintainability of code smells: a large scale empirical investigation. Empir Softw Eng. https://doi.org/10.1007/s10664-017-9535-z
17. Palomba F, Zanoni M, Fontana FA, De Lucia A, Oliveto R (2017) Smells like teen spirit: improving bug prediction performance using the intensity of code smells. In: Proceedings - 2016 IEEE international conference on software maintenance and evolution. ICSME 2016, pp. 244–255. https://doi.org/10.1109/ICSME.2016.27
18. Palomba F, Zanoni M, Fontana FA, Lucia AD, Oliveto R (2019) Toward a smell-aware bug prediction model. IEEE Trans Softw Eng 45(2):194–218. https://doi.org/10.1109/TSE.2017. 2770122
19. Rahman F, Bird C, Devanbu P (2012) Clones: what is that smell? Empir Softw Eng 17(4–5):503–530. https://doi.org/10.1007/s10664-011-9195-3
20. Saboury A, Musavi P, Khomh F, Antoniol G (2017) An empirical study of code smells in JavaScript projects. In: SANER 2017 - 24th IEEE international conference on software analysis, evolution, and reengineering, pp. 294–305. https://doi.org/10.1109/SANER.2017.7884630
21. Shatnawi R, Li W (2006) An investigation of bad smells in object-oriented design. In: Proceedings - third international conference oninformation technology: new generations. ITNG 2006, vol 2006, pp 161–163. https://doi.org/10.1109/ITNG.2006.31
22. Soltanifar B, Akbarinasaji S, Caglayan B, Bener AB, Filiz A, Kramer BM (2016) Software analytics in practice: a defect prediction model using code smells. In: Proceedings of the 20th international database engineering & applications symposium on - IDEAS '16, pp 148–155. https://doi.org/10.1145/2938503.2938553

23. Taba SES, Khomh F, Zou Y, Hassan AE, Nagappan M (2013) Predicting bugs using antipatterns. In: IEEE international conference on software maintenance. ICSM, pp 270–279. https://doi.org/10.1109/ICSM.2013.38

24. Ubayawardana GM, Damith Karunaratna D (2019) Bug prediction model using code smells. In: 2018 18th international conference on advances in ICT for emerging regions (ICTer), pp 70–77. https://doi.org/10.1109/icter.2018.8615550

25. Yamashita A, Moonen L (2013) To what extent can maintenance problems be predicted by code smell detection? -an empirical study. Inf Softw Technol. https://doi.org/10.1016/j.infsof.2013.08.002

26. Zhang X, Zhou Y, Zhu C (2017) An empirical study of the impact of bad designs on defect proneness. In: Proceedings - 2017 annual conference on software analysis, testing and evolution, SATE 2017, vol 2017-Janua, pp 1–9. https://doi.org/10.1109/SATE.2017.9

References

27. Arcelli Fontana F, Mäntylä MV, Zanoni M, Marino A (2016) Comparing and experimenting machine learning techniques for code smell detection. Empir Softw Eng 21(3):1143–1191

28. Azeem MI, Palomba F, Shi L, Wang Q (2019) Machine learning techniques for code smell detection: a systematic literature review and meta-analysis. Inf Softw Technol 108(4), 115–138. https://doi.org/10.1016/j.infsof.2018.12.009

29. Cairo AS, Carneiro GdF, Monteiro MP (2018) The impact of code smells on software bugs: a systematic literature review. Information (Switzerland) 9(11):1–22. https://doi.org/10.3390/info9110273

30. Fowler M, Beck K, Brant J, Opdyke W, Roberts D (1999) Refactoring improving the design of existing code - fowler-beck-brant-opdyke-roberts. Xtemp01

31. Freitas MF, Santos JAM, do Nascimento RS, de Mendonça MG, Rocha-Junior JB, Prates LCL (2016) A systematic review on the code smell effect. J Syst Softw 144(2016):450–477. https://doi.org/10.1016/j.jss.2018.07.035

32. Gradišnik M, Heričko M (2018) Impact of code smells on the rate of defects in software: a literature review. CEUR Work Proc 2217:27–30

33. Kitchenham BA, Budgen D, Brereton P (2015) Evidence-based software engineering and systematic reviews. Chapman & Hall/CRC

34. Le, D., Medvidovic N (2016) Architectural-based speculative analysis to predict bugs in a software system. In: Proceedings of the 38th international conference on software engineering companion, ICSE '16. ACM, New York, NY, USA, pp 807–810. http://doi.acm.org/10.1145/2889160.2889260

35. Ma W, Chen L, Zhou Y, Xu B (2016) Do we have a chance to fix bugs when refactoring code smells? In: Proceedings - 2016 international conference on software analysis, testing and evolution, SATE 2016, pp 24–29. https://doi.org/10.1109/SATE.2016.11

36. Palomba F, Bavota G, Penta MD, Oliveto R, Poshyvanyk D, Lucia AD (2015) Mining version histories for detecting code smells. IEEE Trans Softw Eng 41(5):462–489

37. Palomba F, Nucci DD, Tufano M, Bavota G, Oliveto R, Poshyvanyk D, De Lucia A (2015) Landfill: an open dataset of code smells with public evaluation. In: Proceedings of the 12th working conference on mining software repositories, MSR '15. IEEE Press, Piscataway, NJ, USA, pp 482–485

38. Singh S, Kahlon KS (2011) Effectiveness of encapsulation and object-oriented metrics to refactor code and identify error prone classes using bad smells. ACM SIGSOFT Softw Eng Notes 36(5):1. https://doi.org/10.1145/2020976.2020994

Software Development Artifacts in Large Agile Organizations: A Comparison of Scaling Agile Methods

Ewelina Wińska and Włodzimierz Dąbrowski

Abstract Agile frameworks or methods such as Scrum, Kanban and Extreme Programming are widely adopted and used by small teams within large organizations and start-ups. Established Enterprises, banks, and technology giants are aiming to introduce and implement agile methods that helps organize process in high complexity organizations. Main challenges due to higher complexity relates but are not limited to multiple value streams development pipelines orchestration, communications among many distributed teams, inter team dependencies management, information flow between teams. It can be said that large organizations need to deliver, adapt or introduce new methods to manage development organizations and deliver needed artifacts.

Purpose—This paper aims to compare possible agile frameworks for scaling development organizations working in an agile culture, and their outcomes materialized as artifacts.

Keywords Agile · Agile in large organizations · Scaling Agile organizations Agile software development artifacts · Deliverables in Agile · Scrum · Agile method · Scaling Agile methods · S@S · SAFe · LeSS · Nexus · Artifacts

1 Introduction

It can be said that a major change is in a way how companies approach software development and management of related process and expected outcomes. With the emergence of agile development methods, organizations started to address problems such as delivering values that clients do not need, or a change of requirements during the software implementation process. Changes in product requirements itself,

E. Wińska
Polish Japanese Institute of Computer Science, Warszawa, Poland
e-mail: ewelinawinska@gmail.com

W. Dąbrowski (✉)
Electrical Department, Warsaw University of Technology, Warszawa, Poland
e-mail: w.dabrowski@ee.pw.edu.pl

© Springer Nature Switzerland AG 2020
A. Poniszewska-Marańda et al. (eds.), *Data-Centric Business and Applications*,
Lecture Notes on Data Engineering and Communications Technologies 40,
https://doi.org/10.1007/978-3-030-34706-2_6

101

however, is not the problem, but rather the unnecessary effort that is committed for functionalities that are not need, which results in the waste of resources could be source of resources waste. Companies are moving away from following a waterfall approach to software delivery with its specific implementations such as PRINCE2™ for instance, and are instead incorporating shorter iterations which produce shippable product increments of high-priority features could address decreasing of Time-to-Market value [1]. With strong collaboration between development teams and Product Owners and with frequent feedback cycles, Scrum tends to produce high quality deliverables. Additionally, the inspect and adapt nature of Scrum fosters teams to continually seek better ways of working. With decreased time-to-market and higher quality deliverables also comes increased customer satisfaction. Additionally, frequent Backlog Refinement and reprioritization allows us to more quickly respond to changes, which leads to better customer satisfaction.

Higher team morale and happiness could be achieved with a self-organized team working together with a sense of empowerment. Risks are going to be reduced when small increments are completed in short cycles (not a large amount development work after a significant amount of time and money were spent). We minimize the risk of waste and allow for better flexibility as changes come up. However, as the popularity of the agile approach grows, its applicability is mostly limited to small organizations. Many development organizations have used agile frameworks or methods to work collectively to develop and deliver a software release. However, if more than one development team is working with the same product or project and in the same codebase for a product, they are facing with decreasing individual developer productivity due to integration, communication overhead. If the developers are not on the same collocated team, how will they communicate when they are doing work that will affect each other? If they work on different teams, how will they integrate their work and test the integrated software release?

These challenges appear when two or more development teams are integrating their work into a single shippable software release, and become significantly more difficult when more development teams integrate their work into a software release. A possible solution: scaling methods. Frameworks for managing complex projects in large organizations such as Scaled Agile Framework, Large Scale Scrum, Nexus, Agile Programme Management, Scrum@Scale or Spotify model started to appear [2]. The concept of "artifacts" in Scrum is representing work or value which leads to easier providing of transparency. In addition, artifacts bring opportunities for inspection and adaptation, two core concepts of empirical management models.

Artifacts defined by Scrum and described in Scrum Guide are specifically designed to maximize transparency of key information that is needed for product success. Strong emphasis is also put on organization wide-common understanding of each artifact. This study is focusing on comparison of mentioned frameworks from the defined artifacts perspective. As the complexity of delivery efforts is growing with organization growth, governance models that are incorporating artifacts which capture the results of work being delivered as they are reflecting ideas of transparency, inspection and adaptation are playing a main role.

In this paper authors are focusing on comparing artifacts that have been defined in five possible ways of implementing Scrum in large development organizations: Nexus Framework for scaling Scrum, Large Scale Scrum, Scaled Agile Framework, Agile Programme Management and Scrum@Scale. There are other possible ways of addressing problems during scaling an agile delivery organization such as Spotify Engineering Culture or Site Reliability Engineering framework for Google. Authors are focusing on frameworks that are commonly recognized in the software delivery industry. There are already existing case studies, research papers and publications that enables accurate comparison.

2   Related Work

Scaling agility is not new idea and has been an interest as scaling project demands have raised across the IT sector. During their research Dingsøyr et al. in [3] are proposing unified taxonomy for the purpose of deepening researchers understanding and developing research-based knowledge, needed for in-depth studies.

Rich study of problems relating to self-organizing process implemented by retrospective meetings has been done by researchers in [4]. Study was focused on how a large-scale agile development project uses retrospectives through an analysis of retrospective reports.

During their research published in [5] Thayer, Pyster, and Wood detail twenty problems that many software projects face during development. Their work was published in 1981, but still most issues found on this list are affecting modern projects. For instance, these questions focus on topics such as unclear project requirements, poor project quality, and poor project estimates.

In his study Alqudah and Razali [2], compares what he calls "light" agile methods with "heavy" waterfall-like methods. He points out that heavy approaches favor a "Command-Control" culture, whereas agile approaches tilt toward a culture of "Leadership-Collaboration". The same question of requirements, discussed in [6], is described in terms of "minimal upfront planning" and "comprehensive planning." In his work, Awad attempts to describe the balance between both. It is worth mentioning that in [7], the authors have summarized 12 key lessons on five crucial topics, relevant to large development projects seeking to combine Scrum with traditional project management.

Alqudah and Razali are studying [2] agile methods in large software development organizations with differentiating possible methods due to their maturity. In [8], the authors have a described possible approach to training a team of developers in fields such as Continuous Integration, Continuous Delivery Pipeline, and the Agile methods. As well outcomes of their own observations during such processes.

In [9] authors had conducted detailed analysis about how Agile software development process incorporates 'agile' artifacts such as user stories and product backlogs as well as 'non-agile' artifacts that came from older engineering practices such as extreme programming or management methodologies such as PRINCE2™, for instance designs and test plans.

In [10] authors put emphasis on the problem of how a large organization, informal communication and simple backlogs are not sufficient for the management of requirements and development work. For some large organizations this could be a major problem during agile frame work adoption. In the authors conclusion, there is still little scientific knowledge on requirements management in large-scale agile development organizations.

In [11] authors had approached how nonfunctional requirements are reflected in agile software development, where working software is prioritized over comprehensive documentation. During their research it appears that a focus on "individuals and interaction over processes and tools" encourages minimal documentation.

In [12] researchers are stating that Agile frameworks have become an appealing alternative for companies striving to improve their performance. But what needs to be kept in mind is this creates unique challenges when introducing agile at scale in large delivery organizations. When development teams must synchronize their activities, there might be a need to interface with other organizational units. In their, the paper authors are presenting a systematic literature review on how agile methods and lean software development has been adopted at scale, focusing on reported challenges and success factors in the transformation.

In [13] the author is focusing on the practitioner descriptions of agile method tailoring to explore large scale offshore enterprise development programs with a focus on product owner role tailoring, where the product owner identifies and prioritizes customer requirements.

Authors in [14] are putting effort in to answering why software developers and project managers are struggling to assess the appropriateness of agile processes to their development environments. Their paper identifies limitations that could be applied to many of the published agile processes in terms of the types of projects in which their application may be problematic.

Early adopters of SAFe tend to be large multinational enterprises who report that the adoption of SAFe has led to significant productivity and quality gains. However, little is known about whether these benefits translate to small to medium sized enterprises. In [15] the authors are researching how to adopt, scale and tailor agile methods which depend on several factors such as the size of the organization, business goals, operative model, and needs.

During his research in [16] the author is explaining and clarifying the differences between Waterfall methodologies and Agile development frameworks, which leads to establishing what criteria could be taken into account to properly define project success within the scope of software development projects.

3　Research Method

Research purpose is to address the challenge of artifact transparency in organization wide perspective that are possible within agile frameworks for scaling development organizations working in accordance with agile culture and their outcomes materialized as artifacts. Authors identified this problem while working in commercial projects delivered by large and distributed organizations. Outcomes of presented paper have been implemented and verified in actual projects led by authors, with most emphasis put on the artifacts defined by each reviewed framework. Frameworks that have been chosen for review are Scaled Agile Framework—SAFe, Large Scale Scrum—LeSS, Nexus Framework, Scrum@Scale and Agile Programme Management—AgilePgM.

The aim of this paper is to discuss the comparison of the scaling delivery organization frameworks from a deliverables perspective. During research authors have realized that there is no strict approach to artifacts in scientific papers. Bear in mind that artifacts described in each framework are in fact foundation for transparency parading. Taking this into consideration, the main effort has been put on artifacts introduced in each of frameworks being compared.

Research process has been divided for according steps. Specifying of list of scaling frameworks that will be taken into consideration during the research process. Review of publications about scaling frameworks considered for comparison. Research and review of existing publications related to scaling agile delivery organizations and artifacts defined by specific frameworks that are considered for comparison. Extraction of definitions and implementation guidelines of artifacts defined by each scaling framework are compared in this paper.

Review of different methods has been conducted to identify relevant criteria for comparing and highlighting the differences and similarities in scaling Agile frameworks and artifacts that are being delivered during related process. Based on this authors has conducted review of scientific papers and case studies. Research was based on curated list of publications [1–39] which authors reviewed and took into consideration during analysis.

4　Frameworks that Are Being Compared

The frameworks for scaling delivery teams used in the software industry that authors compared in this study are as follows: Scaled Agile Framework—SAFe, Large Scale Scrum—LeSS, Nexus Framework, Scrum@Scale and Agile Programme Management—AgilePgM. As for this study, the outcomes were compared among these Agile frameworks. Authors had focused on mentioning these five frameworks as they are most recognizable in the software industry and most companies are addressing the challenge of scaling delivery organization using one of them.

4.1   Nexus Framework for Scaling Scrum

Nexus is a framework consisting of roles, events, artifacts, and rules that bind and weave together the work of approximately three to nine Scrum Teams working on a single Product Backlog to build an Integrated Increment that meets a goal. Similarly, in Scrum guide artifacts in Nexus represent work to provide transparency and possibilities for inspection and adaptation loops as well as opportunities for stakeholders to share relevant feedback. In [7] the author is proposing new way of addressing of organizing and orchestrating large development organizations that are using agile frameworks for software delivery.

There is a single Product Backlog for the entire Nexus and all of its Scrum Teams. The Product Owner is responsible for the Product Backlog which translates to managing content, being accountable for its availability, and ordering. Product Backlog view should bind together all dependencies so they can be easily managed and minimalized. Nexus Sprint Backlog is the composite of Product Backlog items from the Sprint Backlogs of the individual Scrum Teams. It is used to highlight dependencies and the flow of work during the Sprint. It is updated at least daily, and best practice is to update it during the Nexus Daily Scrum. The Integrated Increment represents the current sum of all integrated work completed by a Nexus team. The Integrated Increment must be usable and potentially releasable which means it must meet the definition of "Done". The Nexus Integration Team is responsible for a definition of "Done" that can be applied to the Integrated Increment developed each Sprint. All Scrum Teams of a Nexus adhere to this definition of "Done". The Increment is "Done" only when integrated, usable and potentially releasable by the Product Owner.

4.2   Large Scale Scrum—LeSS

Large Scale Scrum framework provides two different large-scale Scrum frameworks. Most of the scaling elements of LeSS are focused on directing the attention of all teams onto the whole product, instead of "my part" approach, global "end-to-end" focus, which are perhaps the dominant problems to solve in scaling. In [18] authors have proposed a detailed way for how to effectively address mentioned problems and have described one of possible ways of implementation scaling delivery organization that is shipping software products in an agile manner. The two frameworks—which are basically single-team Scrum scaled up—are: LeSS for up to eight teams, with the same rule which means each team is up to eight people and LeSS Huge for up to a few thousand people on one product. When an organization is starting a LeSS approach to scaling delivery team, the first thing to clarify is the definition of a products.

The Product Definition needs to be clarified because it affects the scope of the Product Backlog, who will be the Product Owner, the size (in teams) of the Product and hence whether it is a LeSS or LeSS Huge adoption. In [8] authors are also pointing how organization could incorporate different approaches to product management. The suggested way of providing the Product Definition is to conduct a series of questions that concludes in a practical definition. Next artifact that LeSS method is putting emphasis on is the Product Backlog that defines all of the work to be done on the product. Product Backlog Items are not pre-assigned to the teams. The LeSS Product Backlog is the same as in a one-team Scrum environment. During Sprint planning teams are preparing A Sprint Backlog which is the list of work that the teams will need to do for completing the selected Product Backlog Items. The Sprint Backlog is being prepared for each team and there is no difference between a LeSS Sprint backlog and a Scrum Sprint Backlog. Each sprint needs to deliver potentially shippable product increment which translates to a statement about the quality of the software and not about the value or marketability of the software. When a product is potentially shippable, all the work that needs to be done for the currently implemented features has been done and technically the product can be shipped. The Definition of Done is an agreed list of criteria that the software will meet for each Product Backlog Item. Achieving this level of completeness requires the Team to perform a list of tasks. When all tasks are completed, the item is done.

Perfect Definition of Done includes everything that the organization must do to deliver the product to customers. Achieving this should be relatively easy for a one-team product group. When the team can't achieve the perfect Definition of Done then they define 'done' as a subset of the perfect set. As well in [19] authors are suggesting also usage specific tools for sake of successful delivery in large delivery organizations. A crucial part of which is the correct way of incorporating artifacts mentioned above.

4.3 Scaled Agile Framework—SAFe

Scaled Agile Framework for Lean Enterprises is a knowledge base of proven, integrated principles, practices, and competencies for organizations that are approaching Lean, Agile, and DevOps models. More than the sum of its parts, SAFe is a scalable and configurable framework that helps organizations deliver new products, services, and solutions in the shortest sustainable lead time. It's a system that guides the roles, responsibilities, and activities necessary to achieve a sustained, competitive technological advantage. In [1] authors have suggested practical way of implementing such approach and what are needed steps in process that could lead to minimizing waste. Scaled Agile Framework approach is incorporating vast list of artifacts and deliverables.

Major artifacts are: Program Backlog, Program Epics, Program Increment, Program Kanban, Program Level, Program PI Objectives, Roadmap, Team Backlog, Team Demo, Team Kanban, Team PI Objectives, Value Stream Backlog. Short description of major artifacts is as follows. The program backlog is a prioritized list of features intended to address user needs and deliver business benefits. It also includes the enabler features necessary to build the architectural runway. Program epics are initiatives significant enough to require analysis using a lightweight business case and financial approval before implementation. Their scope is limited to a single Agile Release Train and may take several Program Increments to develop. A Program Increment is a timebox in which Agile Release Trains deliver incremental value in the form of working, tested software and systems. Programme increments are typically eight to twelve weeks long, and the most common pattern for a Programme Increment is four development iterations, followed by one Innovation and Planning iteration. Program Increment planning is a cadence-based, face-to-face planning event that serves as the heartbeat of the Agile Release Train, aligning all the teams on the Agile Release Train to a common goal. The program Kanban is a method used to visualize and manage the analysis, prioritization and flow of program epics and features from ideation to completion for a single Agile Release Train. The program level contains the roles and activities needed to continuously deliver solutions via an Agile Release Train. Program Increment objectives are an integrated summary of all the teams' Programme increment objectives for a train. They are used to communicate the plan to stakeholders and to measure accomplishments of the train for a program increment. The roadmap is a schedule of events and milestones that communicate planned deliverables over a timeline. It includes commitments for the planned Program Increment and offers visibility into the forecasted deliverables of the next few PIs. The team backlog contains user and enabler stories that originate from the program backlog, as well as stories that arise locally from the team's specific context. It represents all the things a team needs to do to advance their portion of the system. The team demo is used to measure progress and get feedback at the end of each iteration by demonstrating every story, spike, refactor, and new Nonfunctional Requirements developed in the recent iteration. Team Kanban is a method that facilitates the flow of value by visualizing Work in Progress. This artifact also puts emphasis on establishing Work In Progress limits, measuring throughput, and continuously improving the process. Team Program Increment objectives describe the business and technical goals that an Agile team intends to achieve in the upcoming Programme Increment. They summarize and validate business and technical intent, which enhances communication, alignment, and visibility. The value stream backlog is the repository for all upcoming capabilities and enablers, each of which can span multiple Agile Release Trains and is used to advance the solution and build the architectural runway.

4.4 Agile Programme Management—AgilePgM

The agile programme management philosophy is that an agile programme delivers what is required when it is required—no more no less. This approach is based on five main principles which are listed below. In [20] Dynamic System Development Method Consortium is suggesting these principles could be implemented in an actual environment. Programme goals are clearly and continuously aligned to business strategy. Benefits are realized incrementally and as early as possible. Governance focuses on creating a coherent capability. Decision-making powers are delegated to the lowest possible level. Agile programmers are iterative and could contain both agile and non-agile projects. Agile Programme Management approach is incorporating vast list of artifacts and deliverables.

Major artifacts are: Vision Statement, Business Case, Business Architecture Model, Programme plan, Roadmap, Benefits Realization Plan, Tranche plan, Programme Control Pack, Stakeholder Engagement Strategy, Communications Plan, Governance Strategy, Prioritized Benefits Definition. Short description of major artifacts goes as follow. Vision Statement is primarily a communication tool. The vision keeps the team on track. It also helps with keeping the team on track as the purpose of a vision, particularly in an evolving context. The business case defines the rationale for and expected benefits from the programme and its constituent projects. These should be developed prior to the programme or projects commencing, but in minimal detail. As the programme progresses and the outputs of each project are refined, the benefits and rationale will also be refined. However, it is very important to have enough information up-front to be able to decide to abandon a programme if it does not seem to be delivering on the benefits. Business Architecture Model describes the desired future structure of the organization Programme plan is high level artifact that is composite of two major deliverables Roadmap and Benefits Realization plan. Roadmap, describes the incremental steps and a schedule of tranches for the incremental enablement of consistent capabilities. Benefits Realization plan describes how the benefits identified within the Prioritized Benefits Definition will occur as capabilities are enabled Describes how benefits will be measured. Tranche plan defines the MoSCoW prioritized capabilities projects and costs for the tranche. It also defines the Timeboxes for the projects as subsequent projects and activities that may be subsequent projects or activities. Programme Control Pack contains reports, documents and logs related to the ongoing status of the programme. Stakeholder Engagement Strategy defines approach to be taken towards all groups of stakeholders. Includes either internal and external stakeholders. As well Stakeholder Engagement Strategy should be reviewed and updated often, but at least during Tranche Review. Communications Plan is dimed predominantly at the communication requirements for a given tranche. Each plan will identify how and when key events will be communicated to Stakeholders. Governance Strategy documents and describes the way in which governance will be carried out on the programme. Prioritized Benefits Definition describes benefits predicted to emerge from the change, as well as how benefits realization will be measured. Benefits should be prioritized using MoSCoW rules.

4.5 Scrum@Scale

Scrum@Scale was created to efficiently coordinate an organization where several teams are involved in product delivery (simultaneously working on the same product and effectively the same product backlog). In the same environment it is also a goal to optimize the overall strategy of the organization. Mentioned approach achieves this goal through setting up a "minimum viable bureaucracy" which is possible thanks to scale-free organization governance model. In [17] author had structured and described in detailed way foundations of Scrum@Scale framework. Continuing this concept naturally extends the way a single Scrum team functions across the organization. Scrum@Scale is based on the fundamental principles behind Scrum, Complex Adaptive Systems theory, game theory, and object-oriented technology. To achieve linear scalability effective coordination of multiple teams is needed and crucial. Thanks to its scale-free organization orchestration approach Scrum@Scale is designed to accomplish goal of managing product delivery in large organizations.

With Scrum@Scale core concepts corresponding to original Scrum Guide enterprise artifacts beside Product Backlog, Sprint Backlog and Increment are as follows: Impediment Backlog, Organizational Transformation Backlog, Shared Meta Scrum Backlog, Integrated Potentially Shippable Product Increment. Impediment Backlog is transparent and commonly understood list of impediments that are not possible to resolve on single Scrum Team level and as well inter team dependencies or organizational dependencies. We can describe Organizational Transformation Backlog as a prioritized list of the agile initiatives across whole product delivery organization that need to be accomplished during organizational transformation. Shared Meta Scrum Backlog is Scrum@Scale equivalent of Product Backlog. Similar as product backlog it should be common for the entire organization and should be commonly understood across all Scrum Teams. Integrated Potentially Shippable Product Increment contains shippable increments from all the scrum teams, meets all acceptance criteria and satisfies the entire definition of done (which also includes parts strictly related to integration product as a single shippable integrated increment).

What is also worth mentioning authors of Scrum@Scale are putting emphasis on smaller teams that are proposed in the original Scrum Guide. What is being proposed is to incorporate teams of five, proposition is based on Harvard research that determined optimal team size is four to six people [19]. Experiments with high performing Scrum teams have repeatedly shown that four or five people doing the work is the optimal size. It is essential to linear scalability that this pattern be the same for the number of teams in a Scrum at Scale and beyond this organizational structure.

5   Artifacts Comparison

Table 1 presents the results of the comparison of artifacts corresponding to Backlog for selected frameworks presented in the article. Table content includes artifacts that are proposed and described by compared frameworks. The authors proposed dimensioning of the analyzed methods using a multidimensional method based on the analyses of attributes proposed in the Artifact column. The Description column has a concise description of the validity of the described artifact. For the transparency authors used ⊕to visualize if scaling framework include specific artifact. It

Table 1 Aggregated list of artifacts corresponding to backlog concept that are described in related guidelines for scaled agile framework—SAFe, large scale scrum—LeSS, Nexus framework, Scrum@Scale and Agile programme management—AgilePgM

Artifact	Description	LeSS	Nexus	S@S	SAFe	Agile PgM
Product backlog	Product backlog defines all of the work to be done on the product	⊕	⊕	⊕	⊕	⊕
Sprint backlog	Sprint backlog is the list of work that the team will need to do for completing the selected product backlog items	⊕	⊕	⊕	⊕	⊕
Organizational transformation backlog	A prioritized list of the agile initiatives that need to be accomplished			⊕		
Shared meta scrum backlog	Shared meta scrum backlog is Scrum@Scale equivalent of product backlog. Similar as product backlog it should be common for whole organization and it should be commonly understood across all scrum teams to show dependencies between them			⊕		
Program backlog	The program backlog is the holding area for upcoming features, which are intended to address user needs and deliver business benefits for a single agile release train				⊕	
Nexus sprint backlog	Backlog contains the PBIs that have cross-team dependencies or potential integration issues. In Nexus it is used to highlight dependencies and the flow of work during the sprint		⊕			

appears that only two three artifacts are shared between all five frameworks. Product Backlog, Sprint Backlog and Impediments Backlog however are used in Scrum framework itself which give source for different scaling approaches. Table 2 presents utilizations of integrated increment and programme increment by compared scaling frameworks. Artifacts corresponding to shippable increment assure transparency in what is actual progress of product delivery across organization and indicates real state of product which could be presented to end user. Both Integrated and Programme increments artifacts are encouraging organization to incorporate DevOps culture and could led to sufficient process optimization. Table 3 presents if specifics scaling framework is defining and Health Check checklist and proposing the way implementing it. It is important to state that development organizations should not

Table 2 Aggregated list of artifacts corresponding to product increment

Artifact	Description	LeSS	Nexus	S@S	SAFe	Agile PgM
Integrated increment	The output of every sprint is called an integrated increment. The work of all the teams must be integrated before the end of every sprint—the integration must be done during the sprint	⊕	⊕	⊕		
Program increment	A program increment is a timeboxed event which delivers incremental value in the form of working, tested software and systems. Typically, 8–12 weeks long				⊕	

Table 3 Incorporating of heath check checklist artifact in compared scaling frameworks

Artifact	Description	LeSS	Nexus	S@S	SAFe	Agile PgM
Team health check	Organizational improvement work is very much a guessing game. This artifact is addressing how to assess what needs to be improved, and how will it will be know if it's improving		⊕	⊕	⊕	

forget about sustainability of delivery process which highly relates on team members. Table 4 presents if scaling framework being compared defines Roadmap artifact and its implementation. This artifact relates to ensuring transparency not only for current work perspective but as well what are the future plans among the organization and what could be possible direction in product development. Table 5 presents that each of compared scaling frameworks is defining impediment backlog artifact. This artifact assure transparency in term of inter team communication which is crucial in an environment of big delivery organization.

Table 4 Roadmap artifact definition in compared scaling frameworks

Artifact	Description	LeSS	Nexus	S@S	SAFe	Agile PgM
Roadmap	The roadmap is a schedule of events and milestones that communicate planned Solution deliverables over a planning horizon				⊕	⊕

Table 5 Impediment backlog artifact has been defined in each of compared scaling frameworks

Artifact	Description	LeSS	Nexus	S@S	SAFe	Agile PgM
Impediment backlog	Impediment backlog is transparent and commonly understood list of impediments that are not possible to resolve on single scrum team level and as well inter team dependencies or organizational dependencies	⊕	⊕	⊕	⊕	⊕

6   Evaluation of Compared Methods

Compared methods are answering questions related to scaling agile delivery organizations and how this goal could be achieved. Scrum is an empirical process. For empiricism to work, all artifacts must be transparent. All its artifacts must be transparent to ensure enough accuracy of inspection. All five methods are giving set of best practices that could help with resolving high complexity problems that are incorporating several delivery teams. We can distinguish two main approaches. First based on core Scrum concepts, further developed by Nexus Framework for Scaling Scrum, Large Scale Scrum and Scrum@Scale. Second, which is trying to adapt more agile techniques and as well incorporating waterfall project management approaches such as Scaled Agile Framework and Agile Programme Management. Nexus, S@S and LeSS methods are more organized as frameworks which could serve as canvas for introducing engineering process and other management tools that are helping with achieving business goals. All three of above mentioned methods are putting emphasis on continuous process improvement and self-organization concept. On the other hand, SAFe and AgilePgM are more like process-wide rich methodologies that are focusing on theming process in clear boundaries. Focus is moved into direction of producing strictly defined artifacts which are in most cases related to planning activities or processes. It appears that all discussed frameworks utilize common approach to artifacts. Beside of that AgilePgM is operating on programme level and is not focusing on delivery process or its continuous improvement. Following this both methodologies are duplicating the problem of documentation overhead typical for methodologies that are specific for waterfall approach like PRINCE2^{TM} or PMI, what could lead to not proper implementation of core values of transparency, adaptability and continuous improvement. After analysis of the five above frameworks from the artifact's perspective conclusion appears.

All of them are focusing more on processes and working culture than on actual products. Most of the guidelines are putting emphasis on aspects such as: what meetings should be included in the delivery process, who should cooperate with whom, how to organize single delivery team and whole delivery organization, how to address communication between teams. But in every product delivery initiative what matters most is a working solution that could be shipped to end users and resolves actual problems by generating value. One possible way of embracing transparency could be putting more emphasis on artifact definition and actual usage of defined deliverables. The achievements of development process communicated clearly by artifacts commonly understood among the delivery organization are encouraging inspections and adaptation. What follows with better transparency inspection loops in pace and adaptation having mind delivered products could present more value to end user.

7 Discussion and Further Development

All discussed methods are frameworks that put the most emphasis on cooperation or delivery model. Communication overhead is growing over time, as well as the amount of information related to project. Agile development methods are addressing these issues and are answers for questions about communications management and cooperation among multiple teams. Following this conclusion, those methods should be treated as guidelines and set of high-level best practices, none of those methods provide tools or guidelines for accurate release planning, metrics generation approach or definitions of success factors what could be challenging in non-stop evolving work environment. During the implementation of innovative products in fast changing environment data analytics approach for monitoring and gathering key performance indicators should be introduced and acknowledged as accelerator for initiatives success. Also, when data needed for analysis is changing with tremendous pace big data techniques could be resolutions for many issues related to predicting project outcomes and finding meaningful information. Defining rights metrics should allow an improvement of business outcomes, optimizing waste management and overall optimization of software development life cycle. In next steps we would like to focus on conducting research how big institutions are implementing scaling strategy, what are their problems during the implementation process and what are the tool used for management and change orchestration. We would like to confront this research against the approaches that are being used in startups worldwide. On the other hand authors would like to also put more effort in defining suitable metrics or key performance indicators that will allow accurate tracking of product delivery itself, as well as developments in agile scaling practices.

References

1. West D, Kong P, Bittner K (2017) Nexus framework for scaling scrum, the: continuously delivering an integrated product with multiple scrum teams. Addison-Wesley Professional
2. Alqudah M, Razali R (2016) A review of scaling agile methods in large software development. Int J Adv Sci Eng Inf Technol
3. Dingsøyr T, Mikalsen M, Solem A, Vestues K (2018) Learning in the large—an exploratory study of retrospectives in large scale agile development. XP2018, Porto, Portugal
4. Sorensen R (1995) A comparison of software development methodologies. CrossTalk 8(1):10–13
5. Thayer RH, Pyster AB, Wood RC (1981) Major issues in software engineering project management. IEEE Trans Software Eng 4:333–342
6. Awad M (2005) A comparison between agile and traditional software development methodologies. University of Western Australia
7. Schwaber K (2018) Nexus guide. The definitive guide to scaling scrum with Nexus: the rules of the game. Scrum.org
8. Vodde B, Larman C (2010) Practices for scaling lean & agile development: large, multisite, and offshore product development with large-scale scrum. Addison-Wesley Professional
9. Brinkkemper S, Schneider K (2017) Influence of software product management maturity on usage of artefacts in Agile software development. Product Focused Software Process Improvement PROFES
10. Heikkilä VT, Paasivaara M, Lassenius C, Engblom C (2017) Managing the requirements flow from strategy to release in large-scale agile development: a case study at Ericsson. Empirical Software Engineering, in press
11. Behutiye W, Karhapää P, Costal D, Oivo M, Franch X (2017) Non-functional requirements documentation in Agile software development: challenges and solution proposal. In: International conference on product-focused software process improvement
12. Dikert K, Paasivaara M, Lassenius C (2016) Challenges and success factors for large-scale agile transformations: a systematic literature review. J Syst Softw 119:87–108
13. Bass JM (2015) How product owner teams scale agile methods to large distributed enterprises. Empirical Software Engineering
14. Turk D, France R, Rumpe B (2002) Limitations of Agile software processes. In: Third international conference on extreme programming and flexible processes in software engineering
15. Razzak MA, Noll J, Richardson I, Canna CN, Beecham S (2017) Transition from plan driven to SAFePeriodic team self-assessment. In: Conference: international conference on product-focused software process improvement
16. Veiga AP (2017) Project success in Agile development projects. IT Project Management
17. Sutherland J (2019) The Scrum@Scale guide. The definitive guide to Scrum@Scale: scaling that works. Scrum Inc
18. Vodde B, Larman C (2016) Large-scale scrum: more with LeSS. Addison-Wesley Professional
19. Vodde B, Larman C (2010) Scaling lean & agile development: thinking and organizational tools for large-scale scrum. Addison-Wesley Professional
20. DSDM Consortium (2014) AgilePgM® agile programme management handbook. DSDM Consortium
21. Leffingwell D, Knaster R (2018) SAFe® 4.5 distilled: applying the scaled Agile framework® for lean enterprises. Addison-Wesley Professional

22. Conboy K, Carroll N (2019) Implementing large-scale Agile frameworks: challenges and recommendations. IEEE software—special issue on large-scale Agile development
23. Mazzara M, Naumchev A, Sillitti LSA, Urysov K (2018) Teaching DevOps in corporate environments: an experience report, originally announced July 2018. Published in DEVOPS 2018 workshop
24. Wang Y, Ramadani J, Wagner S (2017) An exploratory study on applying a scrum development process for safety-critical systems. Product Focused Software Process Improvement PROFES
25. Richard HJ (2002) Leading teams: setting the stage for great performances. Harvard Business Press
26. Dingsøyr T, Dybå T, Gjertsen M, Jacobsen AO, Mathisen T-E, Nordfjord JO, Røe K, Strand K (2018) Key lessons from tailoring Agile methods for large-scale software development, originally announced February 2018. Accepted for publication in IEEE IT Professional
27. Scott E, Pfahl D (2017) Exploring the individual project progress of scrum software developers. Product Focused Software Process Improvement PROFES
28. Dobrzyński B, Sosnowski J (2018) Tracing project development in Scrum model. Photonics Applications in Astronomy, Communications, Industry, and High-Energy Physics Experiments
29. Dingsøyr T, Fægri TE, Itkonen J, What is large in large-scale? A taxonomy of scale for Agile software development, for Agile software development. Product-focused software process improvement, vol 8892, pp 273–276. Springer
30. Hohl P, Münch J, Schneider K, Stupperich M (2017) Real-life challenges on Agile software product lines in automotive. Product Focused Software Process Improvement PROFES
31. Svensson RB (2017) Measuring team innovativeness: a multiple case study of Agile and lean software developing companies. Product Focused Software Process Improvement PROFES
32. Klünder J, Hohl P, Fazal-Baqaie M, Krusche S, Küpper S, Linssen O, Prause CR (2017) Reasons for combining Agile and traditional software development approaches in German companies. Product Focused Software Process Improvement PROFES
33. Klünder J, Karras O, Kortum F, Casselt M, Schneider K (2017) Different views on project success: when communication is not the same. Product Focused Software Process Improvement PROFES
34. Ameller D, Farré C, Franch X, Rufian G (2016) A survey on software release planning models. Universitat Politècnica de Catalunya, Barcelona, Spain, Springer International Publishing AG
35. Ram P, Rodriguez P, Oivo M (2018) Software process measurement and related challenges in Agile software development: a multiple case study. In: 19th international conference on product-focused software process improvement (PROFES)
36. Amjad S, Ahmad N, Malik SUR (2018) Calculating completeness of Agile scope in scaled agile development. Published in IEEE Access
37. Matthies C, Kowark T, Richly K, Uflacker M, Plattner H (2016) How surveys, tutors, and software help to assess scrum adoption in a classroom software engineering project. In: Proceedings of the 38th international conference on software engineering companion (ICSE '16)
38. Gren L, Wong A, Kristoffersson E (2018) Choosing agile or plan-driven enterprise resource planning (ERP) implementations—a study on 21 implementations from 20 companies. In: 4th international workshop on the socio-technical perspective in IS development (STPIS2018)
39. Chronis K, Gren L (2016) Agility measurements mismatch: a validation study on three Agile team assessments in software engineering. XP 2016, LNBIP 251, pp 16–27

Tabu Search Algorithm for Vehicle Routing Problem with Time Windows

Joanna Ochelska-Mierzejewska

Abstract Transportation is an important task in the society of today, because economies of the modern world is based on internal and foreign trade. It is obvious that optimized transportation can reduce amount of money spent daily on fuel, equipment and maintenance. Even small improvements can give a huge savings in absolute terms. A popular problem in the field of transportation is Vehicle Routing Problem (VRP). The general vehicle routing heuristics are an important research area as such heuristics are needed for real life problems and each new approach will continue a challenge to make even better heuristics for Vehicle Routing Problems. Tabu Search is still unpopular and rarely used algorithm. Therefore, the paper presents the idea to use Tabu Search algorithm to solve Vehicle Routing Problem with Time Windows constraint. The optimistic and interesting results of using the algorithms for benchmark cases were also presented in the paper. Next, these results were compared with world's best values to show that the implementation of heuristic improved the best known solutions to benchmark cases for many problems.

1 Introduction

Transportation is an important task in the society of today, because economies of the modern world is based on internal and foreign trade. Both services and goods require transportation. Goods must be delivered to customers and services must be provided at the place of supply. It is a very important problem for many financial and ecological reasons. It is obvious that optimized transportation can reduce amount of money spent daily on fuel, equipment and maintenance. Even small improvements can give a huge savings in absolute terms. Several approaches could be taken, one could improve equipment or make the infrastructure better, others could put emphasis on improving existing systems supporting transport processes to improve

J. Ochelska-Mierzejewska (✉)
Institute of Information Technology, University of Technology,
215 Wólczańska Street, 90-924 Łódź, Poland
e-mail: joanna.ochelska-mierzejewska@p.lodz.pl

© Springer Nature Switzerland AG 2020
A. Poniszewsku Maranda et al. (eds,), *Data-Centric Business and Applications*,
Lecture Notes on Data Engineering and Communications Technologies 40,
https://doi.org/10.1007/978-3-030-34706-2_7

decision making. The studies focusing on the second type of improvement are based on simulation, mathematical optimization, econometric methods, data envelopment analysis, neural networks, expert systems, and decision analysis which are used for mathematical modeling of given problem. The use of computerized procedures for planning of the distribution process in the real world brings a mean savings of 5–20% of the transportation costs, so it is definitely worth to analyze such procedures [19, 25].

The transport process is present in all stages of production and distribution of goods and is an important part of their final value (average of 10–20%). Furthermore, transportation takes a great part of the CO_2 pollution in the world today. The transportation sector was in 1998 responsible for 28% of the CO_2 emission, road transportation alone accounted for 84% of the total CO_2 emissions from the transportation sector and it is expected that the CO_2 emissions from the transportation sector is going to increase [19]. Thus, improvements in planning techniques could help in reducing the amount of pollution in the environment caused by transportation. It is so needed that today there are many companies that deal with the production of software for the transportation industry. The use of logistics planning systems becomes effective, thanks to the rapid development of hardware and software platforms in parallel with the increase in computing power. It became possible to achieve the optimum or close to optimum performance within an acceptable time. Increasingly wider availability of auxiliary tools simplifies the creation and maintenance of large systems, which contributes to more widespread use them in practical applications.

A popular problem in the field of transportation is called Vehicle Routing Problem (VRP). The term Vehicle Routing Problem is used to describe a broad class of problems. We are given a fleet of vehicles that often have a common home base, called the depot. Set of customers to be visited and a cost of traveling between each pair of customers and between the depot and each customer is given. The task is to find a route for each vehicle, starting and ending at the depot, such that all customers are served and that the overall cost of the routes are minimized. Problem is very general but typically the solution has to obey several other restrictions, such as capacity of the vehicles or desired visit times at customers. The general vehicle routing heuristics are an important research area as such heuristics are needed for real life problems and each new approach will continue a challenge to make even better heuristics for Vehicle Routing Problems.

Due to the fact that optimization of transportation is still very important problem in the world, it is worth to try to implement a certain solution. Tabu Search is still unpopular and rarely used algorithm. Therefore the main goal of this paper is to implement Tabu Search algorithm to solve Vehicle Routing Problem with Time Windows constraint and to examine obtain results. The presented algorithms has provided excellent results and the implementation is focused on time window constraint.

Section 2 introduces background information on Vehicle Routing Problem. It explain the details of the problem with time windows constraint. Section 3 presents Tabu Search Algorithm and its application to Vehicle Routing Problem. It provides analysis of many important features of presented heuristics, and shows how to apply it in order to obtain better quality solution. Section 4 presents obtained experimental results and their discussion.

2 Vehicle Routing Problem

The last decades have seen an increasing use of computerized procedures for the distribution process. It is possible due to the development of computer systems, and allows to integrate many information systems into the productive and commercial processes. A successful development of modeling and algorithmic tools allows to take into account all the characteristics of models of distribution problems. Algorithms and computer implementations find good solutions for real-world applications within acceptable computing times [25].

In this paper, there is considered only the problem concerning the distribution of goods in a given time period, between depots and final customers by a set of vehicles. Problem is generally known as Vehicle Routing Problems (VRP). VRP is a problem introduced in 1959 by George Dantzig and John Ramser [25]. Authors described a real-world application concerning the delivery of gasoline to gas stations and proposed the first mathematical programming formulation and algorithmic approach for the solution of the problem. In 1964, Clarke and Wright proposed an effective greedy heuristic that improved on the Dantzig-Ramser approach [25]. Since then it proposed a number of sub-problems and different versions of the VRP, and many effective models and optimal and approximate solutions were invented basing on this problem [25].

The goal is to minimize the operation costs, which normally involves reducing the number of vehicles needed or reducing the total distance travelled or time taken. Vehicles are located in depots and perform their movements by using an appropriate road network. Each vehicle perform a route, with start and end at its own depot, all of the customers requirements should be fulfilled, all other constraints should satisfied, and the cost should be minimized. Additionally vehicles do not have to be homogeneous, it can vary in speed or capacity. Many different constraints can be imposed on the routes, vehicles and customers, and the problem is to achieve the optimized solution.

The road network is generally described through a graph, whose arcs represent the road sections and whose vertices correspond to the road junctions and to the depot and customer locations. Graphs can be directed or undirected, depending on whether they can be traversed in only one direction or in both directions, respectively. It is, for example, because of the presence of one-way streets. Each arc is associated with a cost, which generally represents its length. The travel time can be dependent on the vehicle type or on the period during which the arc is traversed. Customer is always represented as vertex of the road, with amount of goods, which must be delivered or collected. The objective of the problem is to find a feasible set of routes for the vehicles so that all requests are fulfilled, and such that the overall travel distance is minimized.

Fig. 1 Vehicle routing
problem—graph
representation

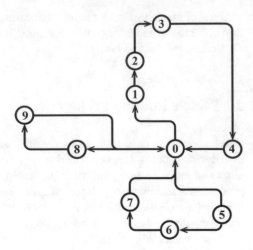

A feasible route of a vehicle should start at the depot, service such a number of clients that the capacity of the vehicle is not exceeded, and finally end at a depot– Fig. 1. The solutions of vehicle and scheduling problems can be used effectively not only for problems concerning the delivery of goods but for the different real-world problems arising in transportation systems as well. Typical applications of this type are, for instance, solid waste collection, street cleaning, school bus routing, transportation of handicapped persons. Sometimes, it is not possible to fully satisfy the demand of each customer. In these cases, the amounts to be delivered or collected can be reduced, or a subset of customers can be left unserved. To deal with these situations, different priorities, or penalties associated with the partial or total lack of service, can be assigned to the customers [18, 25].

2.1 Formal Problem Definition

Vehicle Routing Problem is a problem of graph $G = (V, A)$ where $V = 0, 1, \ldots, n$ is a set of nodes and $A = (i, j)|i, j \in V$ is a set of pair of distinct vertices called arcs or edge. Node $n = 0$ is a depot, other nodes represents customers. Each customer $i \in V' = V \setminus 0$ requires a supply of q products from the depot. A diversified fleet of vehicles is stationed at the depot and is used to supply a customers. The vehicle fleet can contain different vehicle types. Vehicles are available at the depot. Each vehicle has a capacity equal to Q and fixed cost equal to F, that model amortization costs. In addition for each arc $(i, j) \in A$ and each vehicle type k a routing cost, time required to deliver or collect the goods at the customer location, is given and equal to c_{ij}^k. A route is defined as a pair (R, k), where $R = i_1, i_2, \ldots, i_R$ with $i_1 = i_R = 0$ and $i_2 \ldots i_{R-1} \in V'$. It is a simple circuit in G associated with a vehicle k. R is used to refer to the visiting sequence and to the set of customers, including the depot. A route is feasible if the total demand of the customers visited by the route does not exceed vehicle capacity Q. The cost of a route corresponds to the sum of costs of the arcs forming the route plus vehicle fixed cost [2, 4, 13, 25].

$$\sum_{h=1}^{R-1} C_{i_h i_{h+1}}^k + F_k \tag{1}$$

The most general version of the VRP consists of creating a set of feasible routes with minimum total cost, with two demands:

- each customer is visited by exactly one route;
- the number of routes is not greater than number of vehicles.

There are two main objectives in the VRP:

- to find an optimal route to be operated by available vehicles, to supply the customer requirements at minimum cost;
- to find the smallest number of vehicles to supply all customer requirements.

Vehicle Routing Problem can be formulated as presented below. Let $\xi_{ijk} = 1$ if vehicle k visits customer x_j immediately after visiting customer x_i, $\xi_{ijk} = 0$ otherwise. The solution of VRP is to minimize:

$$z = \sum_{i=0}^{N} \sum_{j=0}^{N} \left(C_{ij} \sum_{k=1}^{M} \xi_{ijk} \right) \tag{2}$$

where:

- q quantity of product to be delivered,
- u service time,
- M number of vehicles,
- Q vehicle capacity,
- c cost of path between vertices.

It subjects to:

$$\sum_{i=0}^{N} \sum_{k=1}^{M} = 1, j = 1, \ldots, N \tag{3}$$

It states that a customer must be visited exactly once.

$$\sum_{i=0}^{N} \xi_{ipk} - \sum_{j=0}^{N} \xi_{pjk} = 0, k = 1, \ldots, M, p = 0, \ldots, N \tag{4}$$

It states that if a vehicle visits a customer, it must also depart from it.

$$\sum_{i=1}^{N} \left(q_i \sum_{j=0}^{N} \xi_{ijk} \right) \le Q, k = 1, \ldots, M \tag{5}$$

$$\sum_{i=0}^{N} \sum_{j=0}^{N} C_{ij} \xi_{ijk} + \sum_{i=1}^{N} \left(u_i \sum_{j=0}^{N} \xi_{ijk} \right) \le T, k = 1, \ldots, M \tag{6}$$

Expressions above are the capacity and "cost" limitations on each route.

$$\sum_{j=1}^{N} \xi_{0jk} = 1, k = 1, \ldots, M \tag{7}$$

Expression states that a vehicle must be used exactly once.

$$y_i - y_j + N \sum_{k=1}^{M} \xi_{ijk} \le N - 1, i \ne j = 1, \ldots, N \tag{8}$$

Expression is the elimination condition derived for the travelling salesman problem and which also forces each route to pass through the depot

$$\xi_{ijk} \in 0, 1 \forall i, j, k \tag{9}$$

Above are the integrality conditions. The presented formulation is too complex to be useful in solving VRPs of real-life size [5].

The routes performed to serve customers start and end at one or more depots, located at the vertices of the road graph. Each depot is characterized by the number and types of vehicles associated with it and by the global amount of goods it can deal with. In some real-world applications, the customers are a priori partitioned among the depots, and the vehicles have to return to their home depot at the end of each route. In these cases, the overall VRP can be decomposed into several independent problems, each associated with a different depot.

There are many VRP variances, f.e.:

- Capacitated Vehicle Routing Problem (CVRP) is a VRP in which a fixed fleet of delivery vehicles of uniform capacity must service known customer demands for a single commodity from a common depot at minimum transit cost [14, 25].
- Vehicle Routing Problem with Backhauls (VRPB) is the extension of the CVRP in which the customer set is partitioned into two subsets of linehaul and backhaul customers [20, 25].
- Vehicle Routing Problem with Pickup and Delivery (VRPPD) is a problem where each customer is defined with two quantities: the demand of products to be delivered and picked up at customer, respectively [7, 25].

- The Vehicle Routing Problem with Time Windows (VRPTW) is associated with time windows—an time interval, a period of the day where in the customer should be served [1, 18, 19, 25, 26].

2.2 Vehicle Routing Problem with Time Windows

The Vehicle Routing Problem with Time Windows (VRPTW) is VRP with additional requirements. The VRP consists of designing a set of least cost vehicle routes in such a way that [6, 26]:

- every route starts and ends at the depot;
- every vertex of $V \setminus 0$ is visited exactly once by exactly one vehicle;
- with every vertex is associated a non-negative demand q_i ($q_0 = 0$); the total demand of any vehicle route may not exceed the vehicle capacity Q;
- every vertex requires a service time δ_i ($\delta_0 = 0$); the total length of any route (travel plus service times) may not exceed a given time.

The time window is associated with the customers. It defines a time interval, a period of the day when the customer should be served (for instance, because of specific periods during which the customer is open or the location can be reached, due to traffic limitations). The vehicle should arrive before or within the time window of a customer. Moreover, in case of early arrival, before the start of the time window, it has to wait until the time window opens before service at the customer can start. A solution to VRPTW is a set of tours for a subset of vehicles such that all customers are served only once and time window and capacity constraints are respected.

A formal definition of VRPTW is stated as a graph or network $G = (V, A)$, where $V = 0, 1, 2, \ldots, n$ is a set of vertices and $A = (i, j)|i, j \in V$ is a set of edges. A path P is a sequence of vertices $i_1, i_2, \ldots, i_k$ such that $i_j, i_{j+1} \in A$ for all $1 \leq j \leq k - 1$. The central depot is represented by the nodes 0 and n and vertex i represents customer i for $1 \leq i \leq n$. Moreover:

- c_{ij} – represents the cost of egde, for example travel distance or travel time. For simplicity: $c_{ij} = c_{ji}$;
- b_i – vehicle arrival time at customer i;
- d_i – service time at customer i;
- q_i – demand for customer i;
- e_i – earliest possible arrival time at customer i;
- L – maximum operation time;
- Q – vehicle capacity;
- R_i – a route, such that vehicle will serve customers in order of sequence $i_1, i_2, \ldots, i_k$;
- W_{ij} – waiting time at customer j for vehicle i.

Each vertex has an associated demand, service time and service time window and coordinates. Using coordinates it is possible to calculate distance between nodes and corresponding travel time.

For a given route $R_i = i_1, i_2, \ldots, i_k$ an actual arriving time can be defined as:

$$b_{i_j} = b_{i_{j-1}} + d_{j-1} + c_{i_{j-1}, i_j} \tag{10}$$

which is the arrival time at customer i_{j-1} puls the service time at customer i_{j-1} and travel time from customer i_{j-1} to i_j. It can be modified as:

$$b_{i_j} = \max e_{i_j}, b_{i_{j-1}} + d_{j-1} + c_{i_{j-1}, i_j} \tag{11}$$

which show the earliest arrival time.

The waiting time for vehicle i at customer j is defined as:

$$W_{i_j} = \max 0, e_{i_j} - b_{i_{j-1}} - d_{j-1} - c_{i_{j-1}, i_j} \tag{12}$$

and must satisfy a constraint:

$$b_{i_k} + d_{i_k} + c_{i_k} \leq L \tag{13}$$

R_i is a feasible route. The travel time for this route can be defined as:

$$C(R_i) = c_{0, i_1} + \sum_{j=2}^{k} (c_{i_{j-1}, i_j}) + \sum_{j=2}^{k} W_{i_j} + \sum_{j=2}^{k} d_{i_j} + c_{i_k, 0} \tag{14}$$

A solution S is a collection of routes such that each customer will be covered by exactly one route. The Vehicle Routing Problem with Time Windows can be described as optimization problem of:

$$\min_{s} C(S) = \sum_{i=1}^{l} C(R_i) \tag{15}$$

The objective of VRPTW is to service all customers while minimizing the number of vehicles, travel distance, schedule time and waiting time without changing vehicle capacity and given time windows [2, 25, 26].

3 Tabu Search Algorithm

Very effective heuristic proposed for Vehicle Routing Problem with Time Window is Tabu Search (TS). Computational results on Solomon's benchmarks proved that the proposed TS is comparable in terms of solution quality to the best performing published heuristics. Compared with the many previously defined algorithms that provided the best known, the TS has better performance in exploring irregular search space. Moreover, the proposed algorithm is very good at reducing the required number of vehicles. The proposed algorithm performs well in optimizing most of the 25-customer instances, and has better results for 50-customer instances and 100-customer instances. This proves that Tabu Search is a very powerful technique to solve Vehicle Routing Problem [14, 19].

Tabu Search is a heuristic procedure for solving optimization problems. It uses neighbourhood search to move from the current solution to the best solution. Tabu search algorithm was proposed by Fred Glover [9–12]. He pointed out the controlled randomization in Simulated Annealing to escape from local optima and proposed a deterministic algorithm. In the 1990s, the Tabu Search algorithm became very popular in solving optimization problems in an approximate manner. Nowadays, it is one of the most widespread metaheuristics. This algorithm has obtained very good solutions to many classical and practical problems in such fields as scheduling, telecommunications, neural networks. The particular feature of Tabu Search is a use of memory, which stores information related to the search process [7, 13, 14].

Tabu Search is a a local search based metaheuristics where at each iteration the best solution in the neighborhood of the current solution is selected as a new current solution, even if it leads to increase in the solution cost. Through this mechanism, the method can thus escape from bad local optima. TS behaves like a steepest Local Search algorithm, but it accepts non improving solutions to escape from local optima when all neighbors are non improving solutions. When a local optima is reached, the search carries on by selecting a candidate worse than the current solution. The best solution in the neighborhood is selected as the new current solution even if it is not improving the current solution. Tabu search may be viewed as a dynamic transformation of the neighborhood. Search information can be exploited more thoroughly. To avoid cycling, solutions that were recently examined are forbidden for a number of iterations - tabu restrictions. A short term memory known as Tabu List stores recently visited solutions and gives a possibility to attributes historically found good for searching a better solution. It memorizes the recent search results. At each iteration of TS, the short-term memory is updated. The tabu list may be too restrictive and reject a good move; a non generated solution may be forbidden [14, 25]. Yet for some conditions, called aspiration criteria, tabu solutions may be accepted. The admissible neighbor solutions are those that are non tabu or hold the aspiration criteria. The search stops after a fixed number of iterations or after a number of consecutive iterations have been performed without any improvement to the best solution. The method is still actively researched, and is constantly being improved [7, 13, 14].

There are also two rules concerning the tabu list:

- instead of the same solutions, prohibition includes movements used to produce these solutions;
- a ban on deselecting movement is temporary.

Like tabu restrictions, aspiration criteria can be made time dependent. Aspiration criteria can have significant consequences and should be updated by additional rules [8–12].

3.1 Algorithm

Tabu Search algorithm keeps around a history of recently considered candidate solutions and refuses to return to those candidate solutions until they are sufficiently far in the past. The simplest approach to Tabu Search is to maintain a tabu list L, of some maximum length l, of candidate solutions we have seen so far. Whenever we adopt a new candidate solution, it goes in the tabu list. If the tabu list is too large, we remove the oldest candidate solution and it has no longer a tabu-active state. Proper operation of the algorithm depends on the definition of the following elements [8, 25]:

- objective function,
- Tabu List,
- aspiration criteria,
- final stop condition.

The most important for the proper functioning of algorithm is to define the objective function that allows to generate a move to a new solutions to the problem. Tabu search starts its operation based on the initial solution. Current solution is always replaced by the best solution in the neighborhood, even if this results in moving to worse solution. To stop the possibility of repetition of the same moves and formation of cycles, a neighborhood is defined dynamically with the use of an tabu list, which hold the forbidden moves. Tabu list is used to realize a principle avoid a reverse moves and return moves to already visited places. Additionally it allows to reduce the search space, and thus, reduce computation time. The second principle is to allow selection of the best moves, even if it leads to the worse solution which allows to avoid staying in one local optima. This algorithm involves removing the solution that were already accepted by adding it to collection of tabu moves. In real-life implementations a tabu list have a validity time. It consists of movements made in given number of the last steps. Since a prohibition of movements may prevent a valid solutions from acceptance, additional selection rule was created. Even if a move is included in Tabu List it may be permitted if it fulfills a certain criteria. Aspiration criteria are formulated so that the tabu status of move is overridden when the move leads to a better solution than current one, or when the neighborhood is empty. Aspiration criteria can increase flexibility tabu search by directing research towards more attractive path. Tabu search algorithm uses a short-term memory for storing information about moves that were made. More accurate implementations have also long-term memory. It allows to collect many statistical information, for example, how often victorious moves were carried out. This allows you to modify the motion of the evaluation function, by adding to the original penalty member of the evaluation function as a function of frequency of use of the motion. Stop condition is a condition under which the tabu search is completed. Since the algorithm is open-ended, so the stopping criteria are always needed. In theory, it could go on forever as the optimum is unknown, but in practice, the search has to be stopped at some point. The most commonly used stopping conditions are as follows:

- a fixed number of iterations;
- some number of iterations without an improvement in the objective function value;
- the objective reaches a pre-specified threshold value.

Tabu search method can have a various kinds of extensions and modifications [8, 14, 25].

Below it is presented the steps of Tabu Search [14]:

- Step 1: Choose an initial solution i in S. Set $i' = i$ and $k = 0$.
- Step 2: Set $k = k + 1$ and generate a subset V' of solution in $N(i, k)$ such that either one of the Tabu conditions is violated or at least one of the aspiration conditions holds.
- Step 3: Choose a best j in i' and set $i = j$.
- Step 4: If $f(i) < f(i')$ then set $i' = i$.
- Step 5: Update Tabu and aspiration conditions.
- Step 6: If a stopping condition is met then stop. Else go to Step 2.

3.2 Tabu Search for Vehicle Routing Problem

Tabu Search described by Glover can be used to obtain an optimal solution for VRP. The basic idea is to examine neighbors of a solution and then the best is selected. It prevents cycling by forbidding or penalizing moves previously visited by inserting it in a constantly updated tabu list. Tabu Search operates on the premises that there is no point in accepting a poor solution unless it is to avoid a path already investigated. This provides for moving into new regions of the problems solution space, avoiding local minimum and finally finding the desired solution [6]. There are known many attempts of application Tabu Search algorithm to Vehicle Routing Problem. One of the first attempts is developed by Willard [27]. Here, the problem is first transformed into a TSP by replication of the depot, and the search is restricted to neighbor solutions while satisfying the VRP constraints. Osman in 1991 uses a combination of 2-opt moves, vertex reassignments to different routes, and vertex interchanges between routes [16]. Algorithm developed by Semet and Taillard in 1993 was designed for the solution of a real-life VRP containing several features [21].

Generally Tabu Search algorithm for Vehicle Routing Problem considers a sequence of adjacent solutions obtained by repeatedly moving a random vertex from its current route into another route. Firstly initial solution is constructed. To obtain that customers are assigned randomly to routes. Next a neighbor solution is generated using one of a three popular methods—move type, exchange type or swap type.

Move type:

- choose two routes $R1$ and $R2$ randomly;
- find the smallest distance between two nodes $n1 \in R1$ and $n2 \in R2$;
- relocates the node $n1$ before position of node $n2$,

For example, an element 5 is moved to route $R2$ before element 2:

$$R1 = (015912), R2 = (0210614) \Rightarrow R1 = (01912), R2 = (05210614)$$

Exchange type:

- choose two routes $R1$ and $R2$ randomly;
- find the smallest distance between two nodes $n1 \in R1$ and $n2 \in R2$;
- swap nodes $n1$ and n.

For example, an element 5 and element 2 are exchanged:

$$R1 = (015912), R2 = (0210614) \Rightarrow R1 = (012912), R2 = (0510614)$$

Swap type:

- choose three routes $n1, n2, n3 \in R1$ randomly;
- find all probabilities of swap these three nodes;
- calculate the length of each probability of route $R1$;
- choose the smallest distance of route $R1$.

The procedure to obtain an optimal solution to the VRPTW is to compare neighbor solution with current solution. It can be done by calculating a cost function for both solutions. A cost function is a particular type of objective function that quantifies the optimality of solution. In general view the shorter the route, the lower is the cost value. It can contains additional constraints. For Vehicle Routing Problem with Time Window it can be described as:

$$\sum_{i=0}^{N} \sum_{j=0, i \neq j}^{N} (rt_j - t_i j)p + (t_i j - ct_j)p) \tag{16}$$

where:

- rt represent ready time—moment when vertex is open for service,
- t represents travel time between vertices,
- $t)$ represents closing time—moment when vertex is closed for service,
- p represents penalty coefficient.

To avoid repeating the already examined solution, it is included into Tabu lists— list of solution intent of which was not to prevent a previous move from being reversed. Tabu Search algorithm is open-ended, so the stopping criteria are always needed. For this problem, algorithm will stop searching right after it come to the maximum iteration. Since the optimum is unknown, so the maximum number of iteration is needed [6, 14, 24].

4 Experiments

Solomon Benchmark was used for generation of experimental results. It contains six sets of problem with information about geographical data, vehicle capacity and tightness and positioning of the time windows. The design of these sets highlights several factors that can affect the behavior of routing heuristics. Sets are divided into three groups: randomly generated R, clustered C and semiclustered RC. Semiclustered means that it contains a mix of randomly generated data and clusters. Each dataset contains the 100 customer instances. Sets $R1$, $C1$ and $RC1$ are designed for shorter routes, approximately 5–10 customers per route. Sets $R2$, $C2$ and $RC2$ have a long scheduling horizon for more than 30 to be serviced by the same vehicle. Initial geographical setup is identical for all problems within one type, the only difference in time windows information. A travel time equal the corresponding distance. Solomon benchmark is prepared for two hierarchical objectives: to minimize number of vehicles, and minimize total distance. Solution proposed in this thesis is focused only on minimising the distance, firstly generating a feasible solution [22, 23].

4.1 Experimental Results

This section presents results of experiments. Experiments were taken for each problem sets, datasets from 01 to 08. Each problem was calculated 10 times. Table provides information about calculated distance, cost, travel time. Initial setup which shows geographical positioning of depot and vertices without routes for three groups of data is presented in Fig. 2. After each table a more detailed information for selected iteration is presented. This section shows only selected of the obtained results, odd for $R100$, $C100$, $RC100$ sets and even for $R200$, $C200$, $RC200$.

4.1.1 Random Sets

This section provides post-processed data from random sets received after simulation. Sets $R100$ are configured for maximum 25, 000 iterations and 10 or 12 vehicles number. Sets $R200$ are configured for maximum 25, 000 iterations and 3 vehicles number. Some selected experiment from each set is presented in Table 1 and images of initial and result solution—Fig. 3, Fig. 4, Fig. 5.

Fig. 2 Initial setup for R, C, RC sets

Table 1 Details for selected sets—R101, R107, R208

Solution	Set	Distance	Travel time	Cost
Initial	R101	3542.7185181478503	4678.811994912562	16370.23362807177
Result	R101	1526.5961737636158	2660.035506642823	3585.8664491581126
Initial	R107	3712.509988331251	4861.261531599935	11873.454930904629
Result	R107	1796.4232458955878	2806.1542225950516	542.6615119979629
Initial	R208	3196.2726064829103	4196.27260648291	16312.867748471854
Result	R208	1692.3299201799816	2692.3299201799814	0.0

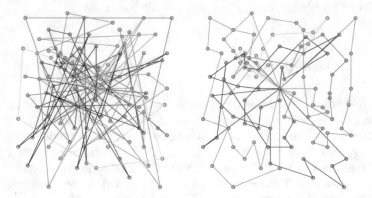

Fig. 3 Initial solution and result for R101

A presented solutions for random sets contains a very interesting information. The most of provided results are good comparing to the best obtained results. Distance value is not very big but it still can be minimized. In many cases a distance and travel time decreased by 50% and cost value decreased by 75%. Some examples are very interesting. Let's consider $R101$ set. A calculated distance is lower than the best result, which suggest that this solution could be better. It is worth to mention that a cost value is big, which means that calculated result is far from being feasible, so it can not be presented as new best result. Calculations obtained from $R107$ are also

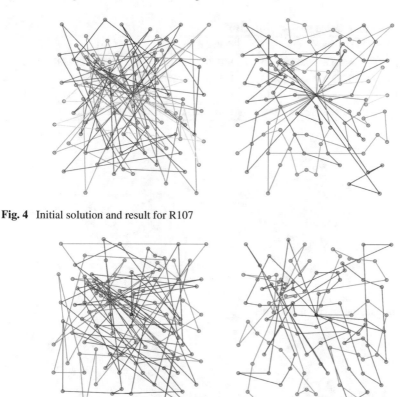

Fig. 4 Initial solution and result for R107

Fig. 5 Initial solution and result for R208

quite good because of low cost value. $R208$ is a set where results after simulations where very good, a cost value happened to be zero, which means that solution is optimal. This set was selected to make more accurate measurements presented in the next subsection.

4.1.2 Clustered Sets

This section provides data from clustered sets obtained after simulation of proposed implementation. Sets $C100$ are configured for maximum $20,000$ iterations and 10 vehicles number. Sets $C200$ are configured for maximum $20,000$ iterations and 3 vehicles number. Results from selected experiment are presented in Table 2 and images of initial and result solution are presented in Figs. — 6 and 7.

A results obtained for clustered sets are not so good comparing to random sets presented in previous subsection. In most cases a calculated cost is big. This means that result is far from optimality. This difference that position of vertices has a very

Table 2 Details for selected sets — C101, C102

Solution	Set	Distance	Travel time	Cost
Initial	C101	4405.863058382725	14076.483297031638	23685.28128233148
Result	C101	2283.4548511303437	11516.656933656921	7176.9559690188
Initial	C102	3988.384366249414	12988.384366249415	75176.47048334271
Result	C102	1963.8588915739913	10964.66513290855	18214.137863353437

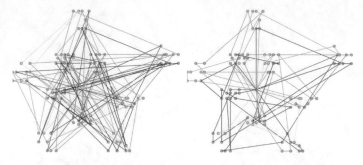

Fig. 6 Initial solution and result for C101

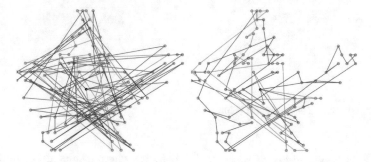

Fig. 7 Initial solution and result for C202

big influence on the result. It is probably an effect of implemented algorithm type. Future work could be focused on implementing different algorithm types, for example exchange type. Most of results has cost between 5, 000 and 9, 000 which is big, and a distance is about 2, 000 worse than the best solution. $C202$ is the worst solution received after simulation. A cost is very big. Provided image and information about values from initial solution proves that algorithm work properly, but a result is not satisfactory in this case.

4.1.3 Random and Clustered Sets

This section presents data from semi-clustered sets obtained after simulation of proposed solution. Sets $RC100$ are configured for maximum 20, 000 iterations and 10 vehicles number. Sets $RC200$ are configured for maximum 20, 000 iterations and 3 vehicles number. Selected results from experiments are presented in Table 3 and images of initial and result solution are presented in Figs. 8 and 9.

Results obtained after simulations of RC sets are good. Calculated cost is not very big, and a distance in many particular examples is close to best obtained distance. Provided images shows that algorithms strives to handle one cluster of vertices by a one path, which is a very good effect. It suggest that with more testing and improve-

Table 3 Details for selected sets — RC202, RC208

Solution	Set	Distance	Travel time	Cost
Initial	RC202	4513.581155095154	5513.581155095154	51136.609531822825
Result	RC202	1990.5435889761923	3018.5607829570263	4948.748015799323
Initial	RC208	4419.555746284099	5419.5557462841	30813.517484689466
Result	RC208	1792.976255787565	2792.9762557875656	25.261166901844604

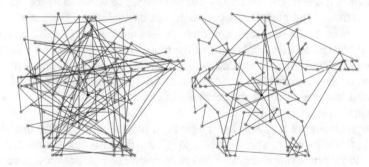

Fig. 8 Initial solution and result for RC202

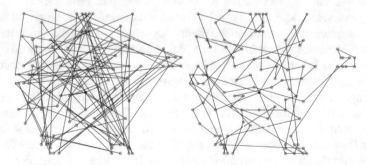

Fig. 9 Initial solution and result for RC208

ments a solution would be a separate routes for each cluster. All solutions preformed a good results with a low cost value with exception of $RC202$ set, which cost is around 7, 000. The best values are presented in table with $RC208$ set. Cost is very low, one of the examples has only 25. RC sets are better than C probably because of group of randomly placed vertices. This allows to easily move nodes from one route to another.

5 Summary

To evaluate the global cost of transportation, a knowledge of the travel cost and the travel time between each pair of customers and between the depots and the customers is required. Road graph can be transformed into a complete graph, where vertices are customers and depots of the road graph. For each pair of vertices of the complete graph, an arc can be defined so the cost is given by the cost of the shortest path between those vertices. Then a travel time can be calculated as a sum of travel times of the arcs belonging to the path. For solving this problem we consider the road graph as associated complete graph, which can be directed or undirected, depending on the property of the corresponding cost and travel-time matrices to be asymmetric or symmetric, respectively. Typical objective that can be considered for the VRP is a minimization of the global transportation cost dependent on the global traveled distance or time. It can be done by minimization of the number of vehicles required to serve all the customers or balancing of the routes, for travel time and vehicle load minimization. There are also applications when vehicle is able to operate more than one route, or the route can last more than one day. There can be also problem with only partial knowledge about the demands of customers or travel costs associated with the arcs of road graph [25].

An large number of metaheuristics have been proposed for solving the VRPTW and most of these metaheuristics have been applied to the Solomon data set. It contains 56 VRPTW instances that all contain 100 customers. The instances contain a variety of customer and time window distributions and have proved to be a challenge for both heuristics and exact methods since their introduction. When the VRPTW has been solved by exact methods one has usually considered minimizing the traveled distance without considering any limits on the number of vehicles. The main objective of the proposed metaheuristics is vehicle minimization, second objective is a travel distance minimization. It is hard to decide which of the previously proposed metaheuristics are the best, as several criteria for comparing the heuristics could be used.

It is important to emphasize that result solution is not always feasible since initial random solution used for simulation is not feasible. Proposed implementation only aspire to obtain correct solution. Non-zero value of cost shows correctness of result - how far from being feasible it is. The results of the experiments confirms that value of cost is significantly reduced after simulation. It additionally provides to lower distance value and travel time. On the other hand, cost value after simulation is still high. It means that optimization of proposed implementation is possible and recommended to improve the performance of a system [3, 15, 17]. Another interesting fact is that presented implementation works better for randomly generated vertices than clustered vertices. It is another problem to be resolved in the future. There is still a place for further research, however it can be said that objective was successfully achieved.

As it was mentioned in previous section there is still a place for further development. Future work includes modification of the algorithm to enable better optimization. It is possible to modify algorithm parameters, or to allow user a better customization. It could also be possible to modify or combine Tabu Search implementation types: move, exchange and swap type. Another interesting approach would be to implement a solution that calculated also vehicle number. Another fact is that our method differs from his method by allowing start solutions not to be feasible. Obtained results are reasonably satisfied but implementing an initial solution to be optimal after creation could provide to much better results. Finally, future development could also provide to solve problem with many different constraints in addition to time windows.

References

1. Beck JC (2003) Vehicle routing and job shop scheduling: what's the difference?. ICAPS03, 267–276
2. Chang Y, Chen L (2007) Solve the vehicle routing problem with time windows via a genetic algorithm. Discret Contin Dyn Syst Suppl 240–249. https://doi.org/10.3934/proc.2007.2007. 240
3. Chomatek L, Poniszewska-Maranda A (2009) Modern approach for building of multi-agent systems, LNCS (LNAI) 5722. Springer, Heidelberg, pp 351–360
4. Christofides N (1976) Vehicle routing problem. Rev Fr D'Automatique 10(V1):55–70
5. Christofides N, Mingozzi A, Toth P (1981) Exact algorithms for the vehicle routing problem, based on spanning tree and shortest path relaxations. Math Program 20(1):255–282
6. Gendreau M, Hertz A, Laporte G (1994) A tabu search heuristic for the vehicle routing problem. Manag Sci 40(10), INFORMS, 1276–1290
7. Gendreau M, Potvin J-Y, Bräumlaysy O, Hasle G, Løkketangen A (2008) Metaheuristics for the vehicle routing problem and its extensions: a categorized bibiography. In: Golden B, Raghavan S, Wasil E (eds) The vehicle routing problem: latest advances and new challenges. Operations research/computer science interfaces, vol 43. Springer, Boston, MA
8. Gendreau M, Potvin J-Y (2010) Handbook of metaheuristics. Springer Science Business Media
9. Glover F (1989) Tabu search: Part 1. ORSA J Comput 1(3):190–206
10. Glover F (1990) Tabu search: Part 2. ORSA J Comput 2(1):4–32
11. Glover F (1990) Tabu search: a tutorial. Interfaces 20(4):74–94
12. Glover F (1995) Tabu search fundamentals and uses
13. Golden BL, Raghavan S, Wasil EA (2008) The vehicle routing problem: latest advances and new challenges. Springer Science and Business Media
14. Hedar A-R, Bakr MA (2014) Three strategies tabu search for vehicle routing problem with time windows. Comput Sci Inf Technol 2(2):108–119. https://doi.org/10.13189/csit.2014.020208
15. Majchrzycka A, Poniszewska-Maranda A (2016) Secure development model for mobile applications, bulletin of the polish academy of sciences. Tech Sci 64(3):495–503
16. Osman IH (1993) Metastrategy simulated annealing and tabu search algorithms for the vehicle routing problems. Ann Oper Res 41:421–451

17. Poniszewska-Maranda A, Majchrzycka A (2016) Access control approach in development of mobile applications. In: Younas M et al (eds) Mobile web and intelligent information systems, MobiWIS 2016, LNCS 9847, Publisher: Springer Heidelberg, pp 149–162
18. Pisinger C, Røpke S (2007) A general heuristic for vehicle routing problems. Comput Oper Res 34(8):2403–2435. https://doi.org/10.1016/j.cor.2005.09.012
19. Ropke S (2005) Heuristic and exact algorithms for vehicle routing problems. Department of Computer Science at the University of Copenhagen
20. Salhi S, Wassan N, Hajarat M (2013) The fleet size and mix vehicle routing problem with backhauls: formulation and set partitioning-based heuristics. Transp Res Part E: Logist Transp Rev 56:22–35. https://doi.org/10.1016/j.tre.2013.05.005
21. Semet F, Taillard E (1993) Solving real-life vehicle routing problems efficiently using tabu search. Ann Oper Res 41:469–488
22. Solomon MM (1987) Algorithms for the vehicle routing and scheduling problems with time window constraints. Oper Res 35(2):254–265
23. Solomon MM (2019). In: VRPTW benchmark problems. http://web.cba.neu.edu/~msolomon/problems.htm
24. Tan KC (2001) Heuristic methods for vehicle routing problem with time windows. Artif Intell Eng 15:281–295
25. Toth P, Vigo D (2002) The vehicle routing problem. Society for Industrial and Applied Mathematics
26. Vacic V, Sobh TM (2004) Vehicle routing problem with time windows. Int Sci J Comput 3(2):72–80
27. Willard JAG (1989) Vehicle routing using r-optimal tabu search. Master's thesis, The management school, Imerial college, London

Code Smell Prediction Employing Machine Learning Meets Emerging Java Language Constructs

Hanna Grodzicka, Arkadiusz Ziobrowski, Zofia Łakomiak, Michał Kawa
and Lech Madeyski

Abstract *Background*: Defining code smell is not a trivial task. Their recognition tends to be highly subjective. Nevertheless some code smells detection tools have been proposed. Other recent approaches incline towards machine learning (ML) techniques to overcome disadvantages of using automatic detection tools. *Objectives:* We aim to develop a research infrastructure and reproduce the process of code smell prediction proposed by Arcelli Fontana et al. We investigate ML algorithms performance for samples including major modern Java language features. Those such as lambdas can shorten the code causing code smell presence not as obvious to detect and thus pose a challenge to both existing code smell detection tools and ML algorithms. *Method*: We extend the study with dataset consisting of 281 Java projects. For driving samples selection we define metrics considering lambdas and method reference, derived using custom JavaParser-based solution. Tagged samples with new constructions are used as an input for the utilized detection techniques. *Results*: Detection rules derived from the best performing algorithms like J48 and JRip incorporate newly introduced metrics. *Conclusions*: Presence of certain new Java language constructs may hide *Long Method* code smell or indicate a *God Class*. On the other hand, their absence or low number can suggest a *Data Class*.

Keywords Code smells detection · Replication study · Machine learning

H. Grodzicka · A. Ziobrowski · Z. Łakomiak · M. Kawa · L. Madeyski (✉)
Faculty of Computer Science and Management,
Wroclaw University of Science and Technology, Wroclaw, Poland
e-mail: lech.madeyski@pwr.edu.pl

H. Grodzicka
e-mail: 226154@student.pwr.edu.pl

A. Ziobrowski
e-mail: 229728@student.pwr.edu.pl

Z. Łakomiak
e-mail: 226190@student.pwr.edu.pl

M. Kawa
e-mail: 228007@student.pwr.edu.pl

© Springer Nature Switzerland AG 2020   137
A. Poniszewska-Marańda et al. (eds.), *Data-Centric Business and Applications*,
Lecture Notes on Data Engineering and Communications Technologies 40,
https://doi.org/10.1007/978-3-030-34706-2_8

1 Introduction

Ever-shifting environment of software development introduces a variety of factors that can affect software quality. Changing requirements, high time pressure and sometimes lack of experience may contribute to creation of code of inferior quality. Code affected by these factors may bring an attentive developer to a conclusion that it "smells". As it is beneficial to be able to grasp the subtleties of such issues, an efficient detection of poor programming habits is a subject of intensive scientific research.

Code smell term was spread by Fowler [3]. He defined smells as something easy to spot (just like real smells) and indicators of a problem since the smells are not inherently bad. Since then, many publications concerning the subject of code smells have been released. A considerable amount of them refer to tools and algorithms for their detection in software projects.

We performed preliminary review of literature on code smell prediction and existing open access datasets of code smells.

One recent publication in this field was published by Arcelli Fontana et al. [1] in the article *Comparing and experimenting machine learning techniques for code smell detection*. In the study attention was paid to four code smells: *Data Class*, *God Class*, *Feature Envy* and *Long Method*. They conduct an extensive comparison of 16 different machine learning algorithms to aid the detection of code smells. They discovered that the best performance was obtained by J48 and Random Forest models. Detection of code smells employing these techniques can provide a very high accuracy—over 96%. There are some premises for imprecision of their research though—these issues will be referred to in the upcoming sections of this paper.

Palomba et al. [10] used the LANDFILL dataset presented in another publication (Palomba et al. [11]) and proposed Historical Information for Smell deTection (HIST) approach exploiting change history information to detect instances of five different code smells (*Divergent Change*, *Shotgun Surgery*, *Parallel Inheritance*, *Blob*, and *Feature Envy*). The results indicate that the precision of HIST is between 72 and 86%, while its recall is between 58 and 100%.

Many papers have introduced a number of tools for detection of code smells. For instance, Arcelli Fontana et al. [1] used iPlasma, PMD, Fluid Tool, Marinescu and Antipattern Scanner. In other publication, Arcelli Fontana et al. [2] additionally used JDeodorant and StenchBlossom to compare tools for code smells detection. Another way to discover presence of smells is *Textual Analysis*, which was described by Palomba [9].

Palomba et al. [11] contributed the dataset containing 243 instances of five smell types from 20 open source projects manually verified by two MSc students. They also presented LANDFILL, a web-based platform aimed to promote the collection and to share code smell datasets.

There is yet another tool for detection of code smells. Designite was created and developed by Sharma [12]. It focuses on C# code, but the author provided a version for Java as well. In our research we have used it to verify its detection capabilities on Qualitas Corpus datasets (Tempero et al. [13]) and within new Java structures. This will be described in one of the following chapters.

Tempero et al. [13] created a web page containing open source Java projects. Arcelli Fontana et al. [1] used 74 projects from this set in order to conduct their research. The biggest drawback of this collection is that the newest version comes from 1 September 2013.

The paper consists of six sections, that are organized in the following manner: in Sect. 2 we describe the reference work and analyze the approach of Arcelli Fontana et al. [1]. Then we discuss the reproduction of Arcelli Fontana et al. [1] experiment. In Sect. 3 we go through the empirical study conducted by our team. In Sect. 4 we present the obtained results which are then augmented by an analysis in the Sect. 5. We conclude the research and indicate future directions in Sect. 6.

2 The Reference Work

Instead of using automated tools for code smell prediction Arcelli Fontana et al. [1] decided on machine learning approach that we aim to reproduce.

2.1 Collecting the Datasets

We began the reproduction of the Arcelli Fontana et al. [1] experiment with obtaining used datasets. Arcelli Fontana et al. [1] sourced datasets from Qualitas Corpus (Tempero et al. [13]), a curated collection of various software systems. As of the date of writing this paper, the current release of Qualitas Corpus is 20130901.

Datasets' last modification dates were retrieved. The results are presented below (Fig. 1).

Arcelli Fontana et al. [1] used Qualitas Corpus distribution from 2012-01-04 which contains source code even from 2002.

State of the art software systems could have completely diverged from design concepts, that were widespread during the creation of such systems. Therefore in the following sections of this paper, we attempted to reason with the usage of such systems as giving reliable results.

We had to filter out Qualitas Corpus systems, which did not appear in the Arcelli Fontana's experiment and augment the source code collection with missing ones. To obtain these results, a set of bash scripts has been written. Individual scripts attempt to tackle problems such as filtering out the incorrect datasets and generating a list of missing systems. We have found that six systems were missing from the newest distribution of the Qualitas Corpus: *freecol, jmeter, jpf, junit, lucene, weka*.

Following systems had differing versions: *freecol, jmeter, junit, lucene* and *weka*, but *jpf* was not present in the Qualitas Corpus newest distribution, despite being attached there in the version, that was used by Arcelli Fontana et al. [1].

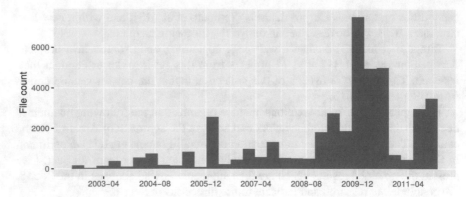

Fig. 1 Histogram presenting last modification dates for the qualitas corpus dataset

Table 1 Summary of
Qualitas Corpus last
modification dates

	Date
Min.	2002-09-02 17:33:14
1st Qu.	2008-08-19 02:07:16
Median	2010-01-03 13:30:36
Mean	2009-05-19 17:18:00
3rd Qu.	2010-08-04 01:26:36
Max.	2011-12-15 21:26:14

The missing systems were manually retrieved from available repositories hosted on Sourceforge or system manufacturer's website. We were successful in obtaining every single one of the systems used in Arcelli Fontana's experiment in the proper version. The archive containing mentioned datasets is available online along with the summary of missing systems (Table 1) prior to their manual supplementation. Additionally, a version of the archive without the test source code has been prepared and shared online likewise.[1]

2.2 Tooling Approach

Arcelli Fontana et al. [1] introduced some inconsistencies in their paper. One of the most vital challenges for the reproduction of the results were choosing the proper version of tools. Arcelli Fontana et al. [1] did not specify advisors nor *weka* versions. Therefore it was needed for us to come up with the most approximate approach.

We decided to choose R language to build the reproduction infrastructure with usage of `caret` and `RWeka` packages, since they are ones of the most popular

---

[1] Datasets—*sources.tar.gz* for full datasets.

tooling[2] choices. As for advisors, we used PMD which date of release was closest to the expected date of creation of Arcelli Fontana et al. [1].

Another downside is the lack of parameter specification, that were obtained by the means of hyperparameter optimization, for the models employed by Arcelli Fontana et al. [1]. Therefore it is not trivial to put the obtained results in the proper context.

2.3 *Caret Versus RWeka Approach*

Both `caret` and `RWeka` [6] R packages have been used during Arcelli Fontana et al. [1] experiment reproduction. Following subsections describe approaches in use of each of the packages.

2.3.1 Caret

Package `caret` (short for *Classification And Regression Training*) is a set of functions that attempt to streamline the process for creating predictive models. The package significantly simplifies data-splitting, pre-processing, feature selection, model tuning using resampling, variable importance estimation and others.

In order to reproduce Arcelli Fontana et al. [1] experiment, the `caret` package has been used. Unfortunately, not all of the models used in the experiment are available in the package, especially `caret` does not include models with the boosting technique AdaBoostM1.

Nevertheless, `caret` has been used because of an ease of cross-validation and grid search (Hsu et al. [7]) parameter estimation technique usage, which has been employed in Arcelli Fontana et al. [1] experiment. Parameters obtained by means of grid search in Arcelli Fontana et al. [1] have not been published in the paper.

Table 2 shows models used from the `caret` package. The rest of algorithms have not been run in `caret` due to limitations of their availability in the package. Parameters presented in the `Tuning parameters` column in Table 2 have been computed with the default `caret` grid search settings. Tune grid used for grid search is derived exclusively for each of the algorithms, based on the model tuning parameters provided by the library, which includes the model.

The granularity in the tuning parameter grid is based on `tuneLength` parameter in the `train` function, which defaults to the division of parameter value space equally on three values. One can specify it manually, but we decided to omit the manual specification of `tuneLength` parameter for sake of performance. Moreover choosing the best tuning parameter is driven by the `best` strategy of `train` function, that selects parameters associated with the best performance in terms of area under the ROC (Receiver Operating Characteristics) curve.

---

[2]https://www.kdnuggets.com/2015/06/top-20-r-machine-learning-packages.html, access: 2019-04-09.

Table 2 Models from `caret`

#	Model	Method value	Type	Library	Tuning parameters
1	C4.5-like Trees	J48	Classification	RWeka	C = 0.01, M = 3
2	Naive Bayes	naive_bayes	Classification	Naivebayes	laplace = 0, usekernel = FALSE, adjust = 1
3	Random Forest	rf	Classification Regression	Random Forest	mtry = 1011
4	Rule-Based Classifier	JRip	Classification	RWeka	NumOpt = 3, NumFolds=3, MinWeights = 2

2.3.2 `RWeka`

Package `RWeka` is an interface to machine learning library `Weka`, a collection of machine learning algorithms for data mining tasks written in Java, containing tools for data pre-processing, classification, regression, clustering, association rules, and visualization (`RWeka` [14], Hall et al. [5]).

In order to reproduce Arcelli Fontana et al. [1] experiment, the `RWeka` package has been used as well. The package allows to use the exact algorithms Arcelli Fontana et al. [1] used, including their boosted versions. Downside of using the package directly is a fact, that it does not provide a simple method for `grid search`, which has been applied as a parameter optimization technique in the experiment we have reproduced.

Table 3 includes models used from `RWeka`. The rest of algorithms have not been run with `RWeka` due to their unavailability in the package.

Table 3 Models from `RWeka`

#	Algorithm name	Default parameters
1	B-J48 Pruned	iterations = 10, C = 0.25, M = 2
2	B-J48 Unpruned	iterations=10, C = 0.25, M = 2
3	B-J48 Reduced Error Pruning	iterations = 10, C = 0.25, M = 2
4	B-JRIP	iterations = 10, F = 3, N = 2.0, O = 2
5	B-SMO RBF kernel	iterations = 10, C = 250007, G = 0.01
6	B-SMO Poly kernel	iterations = 10, C = 250007, E = 1.0
7	J48 Pruned	C = 0.25, M = 2
8	J48 Unpruned	C = 0.25, M = 2
9	J48 Reduced Error Pruning	C = 0.25, M = 2
10	JRIP	F = 3, N = 2.0, O = 2
11	SMO RBF Kernel	C = 250007, G = 0.01
12	SMO Poly Kernel	C = 250007, E = 1.0

Table 4 Missing values overview in Arcelli Fontana et al. [1] datasets

File	No. of missing values
Data-class.arff	75
Feature-envy.arff	92
God-class.arff	76
Long-method.arff	92

2.4 Reproduction Strategy Overview

We attempted to reproduce the results with the approach adequate to Arcelli Fontana et al. [1]. Manually evaluated datasets provided by Arcelli Fontana et al. [1], alongside with available models from `caret` and `RWeka` packages were used.

Firstly, the hyperparameter optimization was done by means of grid search. Method `trainControl` from `caret` employs this method by specifying its `search` parameter as `grid`.

Models were trained using 10-fold cross-validation—for every model there were ten iterations of 10-fold cross-validation, that were later averaged.

ARFF files contain missing values for various metrics—they are marked with explanation mark within a file (Table 4).

Unfortunately, we do not know the strategy to deal with missing values in our reproduction, since Arcelli Fontana et al. [1] does not specify any particular method. It could be therefore crucial to recognize a proper technique to supplement missing values in order to fully reproduce the original results. Based on the trained model, a confusion matrix was computed, thanks to which an accuracy and F-measure were obtained and compared with the results from Arcelli Fontana et al. [1]. The results were augmented with standard deviation and area under ROC as well.

2.5 Classifier Comparison

For reproduction we used ARFF files provided by Arcelli Fontana et al. [1] as an input.

Altogether, 18 algorithms were tested: 16 from RWeka[3] and 6 from Caret.[4] The common classifiers for both libraries are JRip and J48 Unpruned.

Results marked green are those that came out better compared to Arcelli Fontana et al. [1].

In RWeka, for the accuracy the results differ no more than 2% except SMO RBF Kernel (**bold** font) that is unquestionably worse, by 6.53%. Standard deviation for accuracy is always lower for our results—mostly rounds up to zero. In original work it was between 1 and 3 for the given classifiers.

---

[3]https://cran.r-project.org/web/packages/RWeka/index.html, access: 2019-04-09.

[4]http://topepo.github.io/caret/index.html, access: 2019-04-09.

Table 5 RWeka results for *Long Method* (grey) compared with Arcelli Fontana's (white)

#	Classifier	Accuracy	Std dev.	F measure	Std dev.	AUROC	Std dev.
1	B-J48 Pruned	99.20%	0.00	99.40%	0.00	0.9913	0.0083
2	B-J48 Pruned	99.43%	1.36	99.49%	1.00	0.9969	0.0127
3	B-J48 Unpruned	99.41%	0.00	99.56%	0.00	0.9962	0.0042
4	B-J48 Unpruned	99.20%	1.18	99.63%	0.99	0.9969	0.0126
5	B-J48 Reduced Error Pruning	98.99%	0.00	99.24%	0.00	0.9967	0.0027
6	B-J48 Reduced Error Pruning	99.19%	1.31	99.39%	0.87	0.9967	0.0100
7	B-JRip	99.05%	0.00	99.29%	0.00	0.9890	0.0058
8	B-JRip	99.03%	1.26	99.50%	0.94	0.9937	0.0144
9	B-Random Forest	98.99%	0.00	99.25%	0.00	0.9996	2e-04
10	B-Random Forest	99.23%	1.17	99.57%	1.91	0.9998	0.0006
11	B-Naive Bayes	95.52%	0.00	96.57%	0.00	0.9780	0.0035
12	B-Naive Bayes	97.86%	2.37	98.35%	2.02	0.9950	0.0084
13	B-SMO RBF Kernel	95.49%	0.01	96.56%	0.01	0.9900	0.0022
14	B-SMO RBF Kernel	97.00%	2.49	97.75%	2.38	0.9930	0.0116
15	B-SMO Poly Kernel	97.42%	0.00	98.07%	0.00	0.9704	0.0047
16	B-SMO Poly Kernel	98.67%	1.76	99.00%	2.17	0.9852	0.0208
17	J48 Pruned	98.94%	0.00	99.21%	0.00	0.9938	0.0034
18	J48 Pruned	99.10%	1.38	99.32%	1.04	0.9930	0.0151
19	J48 Unpruned	98.92%	0.00	99.19%	0.00	0.9933	0.0035
20	J48 Unpruned	99.05%	1.51	99.28%	1.54	0.9925	0.0168
21	J48 Reduced Error Pruning	98.07%	0.01	98.55%	0.00	0.9887	0.0068
22	J48 Reduced Error Pruning	98.40%	2.02	98.80%	1.13	0.9868	0.0222
23	JRip	98.89%	0.00	99.17%	0.00	0.9880	0.0047
24	JRip	99.02%	1.62	99.26%	1.79	0.9884	0.0181
25	Random Forest	99.23%	0.00	99.42%	0.00	0.9996	1e-04
26	Random Forest	99.18%	1.20	99.54%	1.62	0.9998	0.0011
27	Naive Bayes	93.35%	0.00	94.80%	0.00	0.9649	0.0027
28	Naive Bayes	96.24%	2.39	97.09%	1.72	0.9921	0.0086
29	SMO RBF Kernel	90.44%	0.00	93.29%	0.00	0.8583	0.0066
30	SMO RBF Kernel	97.57%	2.02	98.17%	1.61	0.9732	0.0235
31	SMO Poly Kernel	97.06%	0.00	97.81%	0.00	0.9653	0.0060
32	SMO Poly Kernel	98.67%	1.76	99.00%	1.69	0.9852	0.0208

Table 6 Caret results for *Long Method* (grey) compared with Arcelli Fontana's (white)

#	Classifier	Accuracy	Std dev.	F measure	Std dev.	AUROC	Std dev.
1	SVM C-SVC Linear Kernel	**89.70%**	0.02	**92.62%**	0.01	**0.9225**	0.0228
2	LibSVM C-SVC Linear Kernel	97.31%	2.22	97.97%	1.88	0.9978	0.0044
3	SVM C-SVC Radial Kernel	**71.16%**	0.01	**81.54%**	0.01	**0.7562**	0.0255
4	LibSVM C-SVC Radial Kernel	97.43%	2.09	98.05%	1.32	0.9972	0.0045
5	J48 Unpruned	98.07%	0.01	98.56%	0	0.9842	0.0067
6	J48 Unpruned	99.05%	1.51	99.28%	1.54	0.9925	0.0168
7	Random Forest	98.90%	0	99.17%	0	0.9997	0.0003
8	Random Forest	99.18%	1.2	99.54%	1.62	0.9998	0.0011
9	Naive Bayes	**32.65%**	0	NA	NA	**0.7450**	0.0141
10	Naive Bayes	96.24%	2.39	97.09%	1.72	0.9921	0.0086
11	JRip	98.70%	0	99.03%	0	0.9861	0.0014
12	JRip	99.02%	1.62	99.26%	1.79	0.9884	0.0181

F-measure has similar outcome to accuracy. Our results tend to be nearly the same. The most notable difference is 1.16% excluding SMO RBF Kernel that has lower score by 4.81%. Similarly to the accuracy, F measure standard deviation is rounded up to 0 in nearly every case.

Area under the ROC has very similar results to the original ones. In few cases it is even better. Its standard deviation is approximately two times lower than in Arcelli Fontana et al. [1] research.

The best algorithms from Table 5 are B-J48 and B-JRip, what agrees with Arcelli Fontana's results. All the SMO-based classifiers came out worse in reproduction.

As for caret library, J48 Unpruned, JRip and Random Forest resemble scores achieved by Arcelli Fontana et al. [1]. SVM-based classifiers showed considerable deterioration. Whereas Naive Bayes seems unreliable, it is giving us similarly bad results every time (Table 6).

In conclusion, the reproduction has given consistent results of cross validation for classifiers based on J48, JRip and Random Bayes. SMO and SVM went worse than expected. Naive Bayes gave clearly bad results. The results can be affected by the choice of the libraries and following default behaviour of its components.

Comparison tables for other code smells can by found in Appendix (Sect. 6).

2.6 Learning Curves

Learning curves present the behaviour of models' accuracy with incremental change in number of training examples. Figures 2a, b and 3a, b present results obtained with use of the RWeka package.

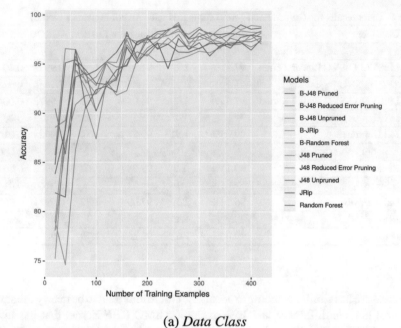

(a) *Data Class*

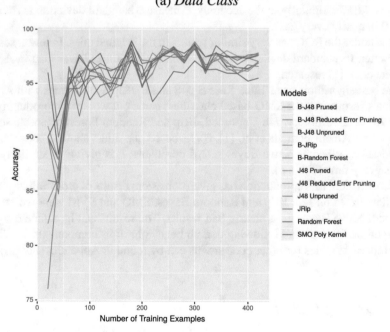

(b) *God Class*

Fig. 2 Learning curves for *Data Class* and *God Class* code smells

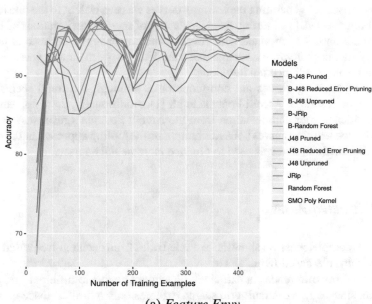

(a) *Feature Envy*

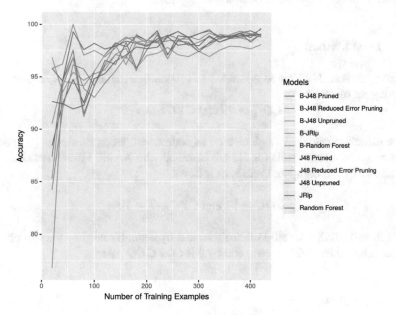

(b) *Long Method*

Fig. 3 Learning curves for *Feature Envy* and *Long Method* code smells

Method used for generating the learning curves was equivalent to the one used by Arcelli Fontana et al. [1]. Starting from the dataset of size 20, ten iterations of 10-fold cross validation has been made with constant increment in the number of training examples after each iteration. Accuracy of a model for a given dataset was computed as a mean from mentioned ten iterations.

Obtained learning curves are non-monotonic, although the trend seems to be similar to the results of Arcelli Fontana et al. [1]. Additionally, there appears to be more significant jitter and deviation from the Arcelli Fontana's results. A deciding factor for such learning curves behaviour might be the tooling approach or the dataset partitioning method, that was used to produce incremental datasets.

2.7 Extracted Rules

For sake of completeness we show the extracted rules from decision-tree based model J48 and from rule-based JRip.

All appearing differences in extracted rules might be due to different grid search approach, since `caret` default grid search explores only a limited space of parameters and no grid search has been employed for `RWeka`.

Metrics' abbreviations can be found in Sect. 3.4.

2.7.1 Long Method

For *Long Method*, J48 Pruned generates a decision tree that can be expressed as the following set of rules:

$$LOC > 79 \text{ and } CYCLO > 9 \tag{1}$$

The detection rule declares a method as code smell, when its size is large and it is complex. Obtained results for J48 are identical with Arcelli Fontana et al. [1].

As for JRip, the computed rule is as follows:

$$LOC \geq 91 \text{ and } CYCLO \geq 10 \tag{2}$$

It differs from the Arcelli Fontana's results by a small margin—we can observe increment by 10 for LOC metric and by 2 for $CYCLO$ metric.

2.7.2 Data Class

For *Data Class* code smell, detection rule derived from the J48 classifier is yet again identical with the results obtained by Arcelli Fontana. The following Boolean logic expression describes the code smell in terms of J48 decision tree:

$$NOAM > 2 \text{ and } WMCNAMM \leq 21 \text{ and } NIM \leq 30 \tag{3}$$

JRip classifier produces the following expression:

$$(WOC \leq 0.352941) \text{ and } (NOAM \geq 3) \text{ and } (RFC \leq 39)$$
$$\text{or } (AMW \leq 1.181818) \text{ and } (NOM \geq 8) \text{ and } (NOAM \geq 4)$$
$$\text{or } (NOM \leq 27) \text{ and } (NOAM \geq 4) \tag{4}$$

The first operand of logical disjunction is nearly the same as Arcelli Fontana's, but unfortunately the rest of logical expression diverges. Obtained results introduce NOM metric and omit $CFNAMM$ and $NOPVA$ metrics.

2.7.3 God Class

Results obtained for *God Class* code smell are convergent for both J48 and JRip. The detection rule of J48 for *God Class* is:

$$WMCNAMM > 47 \tag{5}$$

and for JRip:

$$WMCNAMM \geq 48 \tag{6}$$

Both rules are the same. They are coincident with Arcelli Fontana's results for this code smell.

2.7.4 Feature Envy

Detection rules for *Feature Envy* code smell appear to be the most divergent from the results of Arcelli Fontana et al. [1]. J48 Pruned produces the following decision tree:

$$ATFD > 4 \text{ and } NMO > 8 \text{ and } FDP > 5 \tag{7}$$

This expression introduces FDP and NMO as the metrics that constitute whether a sample source code is influenced by *Feature Envy* code smell. FDP metric describe number of foreign data providers and NMO shows the number of overridden methods in a class. Number of overridden methods provides that class uses data in differences ways. So it could use data from other class than from its own as well. However, Arcelli Fontana's J48 decision tree rule had made use of LAA and NOA metrics.

Same results has been observed for the detection rule extracted from JRip:

$(ATFD \geq 9)$ or $(ATFD \geq 3$ and $LAA \leq 0.)$ and $LOC \geq 20)$

or $(FDP \geq 3$ and $LAA \leq 0.578947)$

$$(8)$$

First operand of the disjunction is identical with Arcelli Fontana's, however the rest of operands differ.

Probable reason of differences in extracted rules is that we used grip search after RWeka computed. Additionally, some rules was extracted by caret functions. Arcelli Fontana et al. [1] used only RWeka.

3 Empirical Study Definition

Arcelli Fontana et al. [1] carried out research on obsolete dataset. We investigated that only fifty are still supported (as of 2019-06-10) (Fig. 4).

3.1 Objective

The aim is to take similar steps to Arcelli Fontana et al. [1] but for more recent and still active projects involving new major Java constructs. We want to find out whether proposed algorithms cope with different dataset and how they perform with new language features. Are they still as effective?

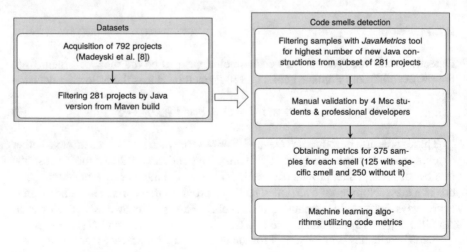

Fig. 4 Steps taken during our research

Fig. 5 Ratio of detected
smells—PMD, Marinescu,
iPlasma, Designite

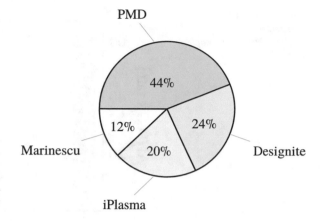

3.2 Extending Code Smell Detection With Designite

In our research we decided to compare results from a new code smell detection tool (advisor) on the same datasets as Arcelli Fontana et al. [1], consisting of 74 system from Qualitas Corpus (Tempero et al. [13]). *Long Method* code smell was exclusively explored, since it is the only one that complies with the newly introduced advisor—Designite for Java Sharma [12].

Our results have been made available online[5] for each of the datasets used by Arcelli Fontana et al. [1].

PMD advisor has detected the highest number of smells (see Fig. 5) and Designite detections coincide the most with Arcelli Fontana's PMD (Fig. 6). One of the factors affecting such result is that PMD and Designite detected the highest number of smells overall. On the other hand, there are notable differences among PMD and Designite comparison and comparisons between following pairs: iPlasma and Designite, Marinescu and Designite. AND/OR ratio for Designite and PMD suggests the closest match between this pair of advisors. However, due to large number of detections for those, the ratio may not be the best accuracy indicator.

3.3 Introducing Datasets Containing Projects with Newer Java Constructions

Despite the fact that the Qualitas Corpus dataset (Tempero et al. [13]) once had been a proper dataset containing Java projects, nowadays it shows several weaknesses such as:

- it is not updated (the current release is 20130901),
- due to lack of updates it contains projects with very old source codes (Table 1),

---

[5]http://madevski.e-informatyka.pl/download/GrodzickaEtA119DataSet.zip.

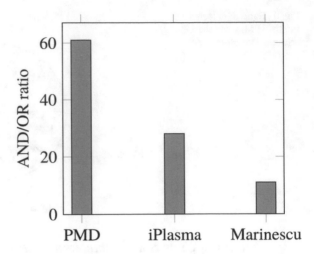

Fig. 6 AND/OR ratio of detected smells—Designite compared with other advisors

- old source codes mean that new Java constructions cannot be observed nor examined.

For a solution to these problems, we decided to introduce a new dataset containing 792 projects Madeyski et al. [8]. The main aim of creating the dataset is to study the impact of newer Java constructions on code smells detection. We concluded, that the current breakthrough Java version is the Java 8. Java 8 has been chosen as the determinant of the freshness of the projects, due to the multitude of new constructions it introduces and its long-term support (LTS) status.

3.3.1 Information About The new Dataset

All of the projects in the base dataset Madeyski et al. [8] are open source and available on GitHub. Not all of them are plain Java projects and, additionally, not necessarily use at least Java 8. For this reason the base dataset had to go through filtering.

The R package `reproducer` [8] contains basic information to obtain the projects in their specific versions from GitHub and had been used while collecting them. Further description can be found in the Appendix [4].

The dataset contains actively developed projects (see Table 7, Fig. 7). A substantial difference can be observed comparing the datasets' summary with the Qualitas Corpus datasets' summary (see Table 1).

3.3.2 Filtering Java Maven Projects from The base Dataset

In order to ensure that projects comprise Java, we decided to filter them upon Maven build tool. Using Maven build information allowed Java version retrieval aside. Although the base dataset contains constantly updated projects, we were particularly interested in ones with the Java version equal to or higher than the Java 8 version.

Table 7 Summary of last modification dates of 792 projects from the new dataset

	Date
Min.	2002-02-28 13:58:11
1st Qu.	2016-06-18 19:45:16
Median	2018-03-06 08:29:50
Mean	2017-03-21 00:46:19
3rd Qu.	2018-12-04 15:49:56
Max.	2019-04-16 22:55:47

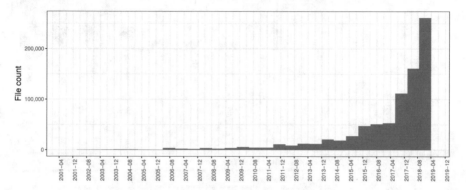

Fig. 7 Overview of last modification dates of 792 projects from the new dataset

3.3.3   Checking Java Version In the Maven Build Tool

The first step to obtain new constructions is filtering projects' build tools by Java version used in builds. Figure 8, Table 8 show distribution of build tools among projects.

There are 439 plain Maven projects, others combine different or multiple build tools. Projects marked as *other or lack of* indicate use of other build scripts and tools than *pom.xml* for Maven, *build.gradle* for Gradle or *build.xml* for Ant.

Additionally, *build.properties* and *.travis.yml* files have been searched for Java versions.

The Maven build tool, as the most frequently used, has been selected to retrieve information about the Java version from Maven-based projects. Maven configuration file (`pom.xml`) has been searched for Java version used during the projects' builds. Extracting the version from Maven build script was not always straightforward because of multiple ways of encoding it. Moreover, non-Maven projects or these with unconventionally specified build were omitted and Java version for them was marked as *unknown* (Fig. 9, Table 9).

After filtering the 792 projects for at least Java 8 version, the following results were obtained (Table 12, Fig. 12).

Fig. 8 Projects' build tools

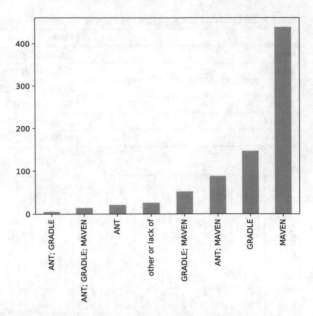

Table 8 Projects' build tools

Build tool(s)	Number of projects
Maven	439
Gradle	147
Ant & Maven	88
Gradle & Maven	52
Ant	21
Ant & Gradle & Maven	14
Ant & Gradle	5
other or lack of	26

Fig. 9 Projects' Java versions

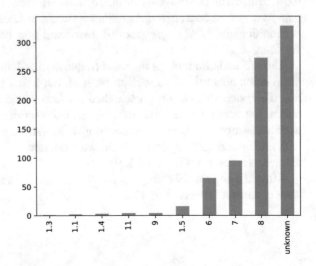

Table 9 Projects' Java versions

Java version	Number of projects
8	273
7	95
6	65
1.5	16
11	4
9	4
1.4	3
1.1	2
1.3	1
unknown	329

Table 10 Summary of last modification dates of 281 projects from the new dataset

	Date
Min.	2005-02-03 22:37:19
1st Qu.	2017-08-28 17:23:51
Median	2018-08-02 11:18:04
Mean	2017-10-28 07:30:51
3rd Qu.	2019-01-03 05:27:38
Max.	2019-04-16 18:01:42

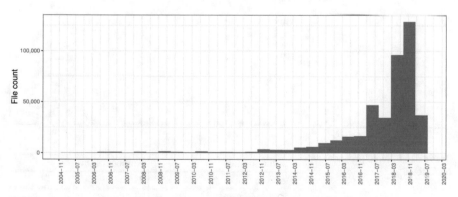

Fig. 10 Overview of last modification dates of 281 projects from the new dataset

As for summary of original datasets' subset, following information can be observed (Table 10, Fig. 10).

Appendix [4] contains details required to obtain the filtered projects in their specific versions from GitHub. Overview of build tools and Java versions used in the obtained projects is shown on Figs. 11, 12 and Tables 11, 12.

Fig. 11 Filtered projects' build tools

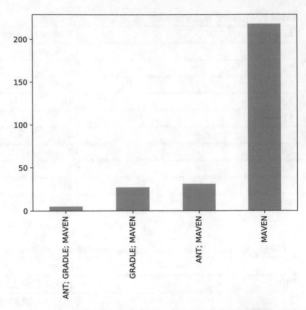

Fig. 12 Filtered projects' Java versions

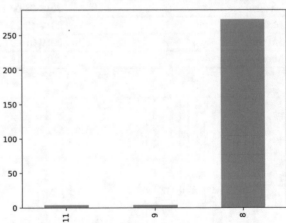

Table 11 Filtered projects' build tools

Build tool(s)	Number of projects
Maven	218
Ant & Maven	31
Gradle & Maven	27
Ant & Gradle & Maven	5

Table 12 Filtered projects' Java versions

Java version	Number of projects
8	273
11	4
9	4

3.4 Filtering And retrieving Information About the Actual Use Of new Java Constructions Using The JavaParser

JavaMetrics is our custom solution that employs JavaParser[6] to automate metric derivation. It exposes the following metrics that can be run against the provided project:

- Access to Foreign Data ($ATFD$)
- Cyclomatic Complexity ($CYCLO$)
- Foreign Data Providers (FDP)
- Locality of Attribute Access (LAA)
- **Lambda Density (LD)**
- Lines of Code (LOC)
- **Method Reference Density (MRD)**
- Number of Inherited Methods (NIM)
- Number of Accessor Methods ($NOAM$)
- **Number of Lambdas (NOL)**
- Number of Methods (NOM)
- Number of Mutator Methods ($NOMM$)
- **Number of Method Reference ($NOMR$)**
- Number of Public Attributes ($NOPA$)
- Number of Private Attributes ($NOPV$)
- Weighted Method Count (WMC)
- Weighted Method Count of Not Accessor or Mutator Methods ($WMCNAMM$)
- Weight of Class (WOC)

Custom metrics are written in **bold**. Both LD and MRD refer to constructions (lambda and method reference) density in class, for instance $LD = NOL/LOC * 100$ (result in percent). NOL and $NOMR$ are implemented for methods separately to use them for *Long Method* detection.

JavaMetrics utilizes visiting the Abstract Syntax Tree (AST) nodes in deriving the metrics. JavaParser allows to traverse the AST with callback invocation only for certain types of nodes. Thanks to robust JavaParser API, it is possible to detect Java language constructs and process them accordingly.

---

[6]https://javaparser.org, access: 2019-06-10.

The tool is flexible enough to implement other custom metrics without changing the existing code structure (open-closed principle). A CSV file is generated as the result of the *JavaMetrics* run. The output consists of parsed unit details such as package, class, method signature and additional information whether methods and fields are final or static. Moreover it provides the selected metrics values as the succeeding columns in the output.

Unfortunately, even though the metrics derivation works as expected, the ergonomics of the JavaParser may pose a challenge to completely automate metric derivation process. Missing part involves the resolution of classes. For example the NIM metric expects information about the superclass that is most likely in the separate file. JavaParser tool has incorporated a second tool called JavaSymbolSolver as a part of the library. It allows to resolve classes by means of the `import` clauses and referenced files. Both the directory containing the packages under the resolution and external dependencies in the form of JAR files shall be provided.

In order for *JavaMetrics* to work seamlessly, we would have to parse the provided directory to find the directories enclosing the packages and parse the build automation tool files (such as *pom.xml* for Maven) to identify the external dependencies. Those external dependencies would have to be downloaded and appropriately registered within the *JavaMetrics*. This issue is the matter of the future directions of work.

3.5 Manual Tagging

Our tool named *JavaMetrics* was used for initial filtering of samples from dataset (Madeyski et al. [8]) with the highest number of lambdas and method references. *JavaMetrics* let us conduct in-depth analysis of the selected samples by bringing metrics.

Code smell recognition was performed mostly by us, 4th year Software Engineering students and junior developers with approximately one year of professional experience. We were validating selected samples till reaching 125 *Long Methods*. Our judgement was partly based on metrics from *JavaMetrics* that gave us overview for each sample.

Tagged samples have been replenished with some tagged by professional developers (Table 13) mentioned in Acknowledgement following Sect. 6. Due to similarities in their specification, we have treated *Blob* as *God Class* for the purpose of our research.

Eventually 125 samples were selected for *Long Method*, *Data Class* and *God Class* along with complementary 250 samples without those smells. All of professionally tagged samples were used. Such datasets (with resulting metrics) were used as an input for the machine learning algorithms.

Table 13 Tagged samples from professional developers

Code smell	Minor or higher severity	Nonsevere
Long Method	0	3
Data Class	5	26
Blob	15	16

4 Results

Results of manual tagging consist of 375 samples—125 samples with specific code smell and 250 without it.

Each 375 samples along with their metrics were used as an input to build code smell prediction rules employing machine learning algorithms.

4.1 Classifier Performance

Using algorithms from Caret that gave best results in Sect. 2.5 we were able to test them for three code smells on the new data we gathered (Tables 14, 15, 16).

Table 14 caret results for *Long Method*

#	Classifier	Accuracy	Std dev.	F measure	Std dev.	AUROC	Std dev.
1	J48	96.1 %	NA	97.08 %	NA	0.955	NA
2	rf	96.81 %	NA	97.63 %	NA	0.9963	NA
3	naive_bayes	94.68 %	NA	96 %	NA	0.9778	NA
4	JRip	96.81 %	NA	97.61 %	NA	0.9632	NA

Table 15 caret results for *God Class*

#	Classifier	Accuracy	Std dev.	F measure	Std dev.	AUROC	Std dev.
1	J48	89.01 %	NA	91.46 %	NA	0.8947	NA
2	rf	86.52 %	NA	89.73 %	NA	0.9462	NA
3	naive_bayes	83.69 %	NA	88.32 %	NA	0.9135	NA
4	JRip	86.88 %	NA	89.52 %	NA	0.903	NA

Table 16 caret results for *Data Class*

#	Classifier	Accuracy	Std dev.	F measure	Std dev.	AUROC	Std dev.
1	J48	97.13 %	NA	93.54 %	NA	0.9036	NA
2	rf	93.62 %	NA	95.24 %	NA	0.976	NA
3	naive_bayes	89.72 %	NA	92.14 %	NA	0.9436	NA
4	JRip	92.2 %	NA	94.18 %	NA	0.9177	NA

4.2   Extracted Rules

The J48 and JRip algorithms provide human readable detection rules as the part of the output. We present them to show significant metrics used for certain code smells detection.

4.2.1   Long Method

For *Long Method*, J48 generated a decision tree that can be expressed as:

$$LOC\_M > 51 \tag{9}$$

JRip classifier computed the following rule:

$$LOC\_M \geq 52 \tag{10}$$

4.2.2   God Class

J48 generated a decision tree that can be expressed as following set of rules:

$(WMCNAMM \leq 49$ and $WMCNAMM > 8$ and $LD \leq 2.028986$ and $NOM \leq 14)$
or $(WMCNAMM \leq 49$ and $WMCNAMM > 32$ and $LD \leq 2.028986$ and $NOM > 14$ and $MRD > 0.470958)$
or $(WMCNAMM > 49$ and $NOL\_C > 13)$
or $(WMCNAMM > 49$ and $NOL\_C \leq 5)$
or $(WMCNAMM > 49$ and $NOL\_C \leq 6$ and $NOL\_C > 5$ and $NOMM \leq 2$ and $LD > 1.595745)$
or $(WMCNAMM > 49$ and $NOL\_C \leq 13$ and $NOL\_C > 6$ and $NOMM \leq 0)$
or $(WMCNAMM > 49$ and $NOL\_C \leq 13$ and $NOL\_C > 6$ and $NOMM \leq 2$ and $NOMM > 0$ and $LD \leq 1.647059)$
or $(WMCNAMM > 49$ and $NOL\_C > 13)$

$$\text{or } (WMCNAMM > 49 \text{ and } NOL\_C \leq 13 \text{ and } NOL\_C > 5 \text{ and } NOMM > 2) \tag{11}$$

As for JRip, the detection rule is as follows:

$$(WMCNAMM \geq 51) \text{ or } (NOL\_C \leq 5 \text{ and } WMCNAMM \geq 32) \tag{12}$$

4.2.3   Data Class

The detection rule derived from J48 can be expressed as:

$(NOL\_C \leq 6$ and $WOC \leq 0.769231)$
or $(NOL\_C \leq 6$ and $WOC > 0.769231$ and $NOPA \leq 3$ and $NOMM > 1)$

$$\text{or } (NOL\_C \leq 6 \text{ and } WOC > 0.769231 \text{ and } NOPA > 3) \tag{13}$$

JRip classifier produces the following expression:

$$(NOL\_C \leq 6 \text{ and } WOC \leq 0.76) \text{ or } (LD \leq 2.083333 \text{ and } NOPA \geq 4) \quad (14)$$

4.3 Learning Curves

Measuring accuracy while incrementing number of training examples provided us with learning curves for each code smell (Fig. 13).

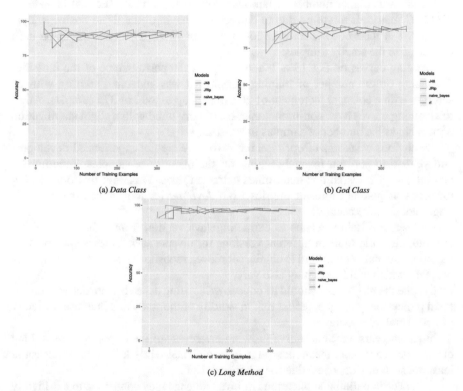

(a) *Data Class*   (b) *God Class*

(c) *Long Method*

Fig. 13 Learning curves for code smells

5 Discussion

During manual tagging we have found that *Long Method* code smell occurs approximately two times less frequently than *God Class*. This might be due to new constructions contributing to shorten overall code length.

J48 for *God Class* (Eq. 11) shows that high lambda count and high $WMCNAMM$ metric values with $CYCLO$ as the embedded metric may indicate a *God Class*. Lambdas concise syntax contributes to the class complexity, since more operations can be expressed with less code.

Data Classes usually incorporate no logic and this is noticeable in their equation which discards higher number of lambdas. JRip detection rule (Eq. 14) introduces NOL_C metric. It might be due to data classes having little to no business logic embedded in their code. Therefore low values of NOL_C metric might be used to recognize the smelly samples.

Learning curves on Fig. 13 a, b and c depicts the performance of the classifier depending on the number of training samples. It was generated starting with 20 samples with a constant increment of 20 samples up to total of 375 samples. All of the learning curves have non-monotonic, oscillatory trend with a slight stabilization noticeable as the number of samples increases.

An obvious weakness of our research is that we are not experienced developers and yet had to perform manual tagging for the majority of samples. Tagging is a crucial part of research that determines extracted rules. We admit that our tagging process was guided by result metrics from *JavaMetrics* tool, hence the outcome might not be quite accurate.

Another uncertainty relates to extracting Java version from build tools. There are multiple build tools in different versions for diverse Java releases specification methods that makes them hard to obtain. Moreover, repositories often consist of many projects, which build tools may also vary.

Despite the results, one should be concerned as using datasets from actively developed projects may carry potential risk in validity of the study and increase difficulty of code smell prediction.

In recent years we can observe growing awareness of the importance of writing clean code. Thus more recent dataset may bring consistently less code smells (or at least not such distinct ones) due to its better design.

Advances in building applications in Java language may contribute to a difficulty in detection of certain code smells. Using various libraries and frameworks can affect the detection of code smells as well. As an example, Lombok[7] can completely exclude presumable *Data Classes* from the detected instances due to a slight decrease in WOC metric. Therefore our approach may not be suitable for every project and shall not be taken as a generic way to predict code smells.

---

[7]https://projectlombok.org, access: 2019-06-12.

6 Conclusions and Future Directions

Using lambda constructs contributes naturally to more concise syntax which may obfuscate certain code smells like *Long Method*. Moreover, using Java Streams does not affect the increase of $CYCLO$ metric, which may further distort the predictions for code smells indicated by such metric.

In addition to extending the tool with more metrics concerning emerging Java language constructs and creating the external dependency parsing approach, we have identified the possible future directions of research and development:

- **Tagging**. Employing professional developers for manual tagging.
- **Algorithms**. Involving more ML algorithms in creating predictive models.
- **Metrics**. Extend *JavaMetrics* tool to correctly parse external dependencies.
- **Constructions**. Considering wider range of modern Java features than lambda expressions and method reference.

Acknowledgements This work has been conducted as a part of research and development project POIR.01.01.01-00-0792/16 supported by the National Centre for Research and Development (NCBiR). We would like to thank Tomasz Lewowski, Tomasz Korzeniowski, Marek Skrajnowski and the entire team from *code quest sp. z o.o.* for tagging code smells and for all of the comments and feedback from the real-world software engineering environment.

Appendix: Reproduction classifier comparison

Following tables present comparison of our reproduction results to Arcelli Fontana et al. [1] (Tables 17, 18, 19).

Table 17 RWeka results for *Data Class* (grey) compared with Arcelli Fontana's (white)

#	Classifier	Accuracy	Std dev.	F measure	Std dev.	AUROC	Std dev.
1	B-J48 Pruned	98.78%	0.00	99.05%	0.00	0.9910	0.0050
2	B-J48 Pruned	99.02%	1.51	99.26%	1.15	0.9985	0.0064
3	B-J48 Unpruned	97.55%	0.01	98.10%	0.01	0.9856	0.0073
4	B-J48 Unpruned	98.67%	1.79	98.99%	1.86	0.9984	0.0064
5	B-J48 Reduced Error Pruning	98.09%	0.01	98.52%	0.01	0.9930	0.0045
6	B-J48 Reduced Error Pruning	98.07%	2.47	98.55%	1.36	0.9951	0.0103
7	B-JRip	98.55%	0.00	98.87%	0.00	0.9950	0.0037
8	B-JRip	98.83%	1.60	99.12%	1.21	0.9959	0.0111
9	B-Random Forest	97.55%	0.00	98.12%	0.00	0.9988	4e-04
10	B-Random Forest	98.57%	1.68	98.94%	2.08	0.9993	0.0017
11	B-Naive Bayes	86.17%	0.01	88.34%	0.01	0.9221	0.0117
12	B-Naive Bayes	97.33%	2.29	98.02%	2.01	0.9955	0.0091
13	B-SMO RBF Kernel	94.21%	0.01	95.48%	0.00	0.9848	0.0033
14	B-SMO RBF Kernel	97.14%	2.61	97.86%	2.11	0.9782	0.0286
15	B-SMO Poly Kernel	94.34%	0.00	95.59%	0.00	0.9387	0.0036
16	B-SMO Poly Kernel	96.90%	2.25	97.65%	2.35	0.9862	0.0173
17	J48 Pruned	98.47%	0.01	98.81%	0.00	0.9834	0.0076
18	J48 Pruned	98.55%	1.84	98.91%	1.39	0.9864	0.0219
19	J48 Unpruned	98.52%	0.00	98.85%	0.00	0.9816	0.0043
20	J48 Unpruned	98.38%	1.87	98.79%	1.90	0.9873	0.0208
21	J48 Reduced Error Pruning	96.86%	0.01	97.55%	0.01	0.9795	0.0081
22	J48 Reduced Error Pruning	97.98%	2.46	98.46%	1.40	0.9839	0.0211
23	JRip	97.5%	0.01	98.05%	0.01	0.9782	0.0089
24	JRip	98.17%	2.18	98.62%	2.07	0.9809	0.0241
25	Random Forest	97.55%	0.00	98.12%	0.00	0.9989	5e-04
26	Random Forest	98.95%	1.51	99.29%	2.05	0.9996	0.0014
27	Naive Bayes	79.69%	0.01	81.99%	0.01	0.9475	0.0063
28	Naive Bayes	96.12%	2.95	97.04%	1.95	0.9938	0.0099
29	SMO RBF Kernel	95.43%	0.00	96.54%	0.00	0.9394	0.0039
30	SMO RBF Kernel	97.05%	2.38	97.78%	2.11	0.9686	0.0286
31	SMO Poly Kernel	94.23%	0.00	95.51%	0.00	0.9379	0.0031
32	SMO Poly Kernel	96.60%	2.76	97.41%	2.13	0.9912	0.0138

Table 18 RWeka results for *Feature Envy* (grey) compared with Arcelli Fontana's (white)

#	Classifier	Accuracy	Std dev.	F measure	Std dev.	AUROC
1	B-J48 Pruned	96.01%	0.01	97.01%	0.00	0.9880
2	B-J48 Pruned	96.62%	2.78	97.41%	2.16	0.9900
3	B-J48 Unpruned	93.79%	0.01	95.39%	0.01	0.9684
4	B-J48 Unpruned	96.50%	2.96	97.37%	2.37	0.9899
5	B-J48 Reduced Error Pruning	95.28%	0.01	96.48%	0.01	0.9829
6	B-J48 Reduced Error Pruning	95.90%	3.11	96.90%	2.24	0.9866
7	B-JRip	96.19%	0.01	97.15%	0.00	0.9869
8	B-JRip	96.64%	2.84	97.44%	2.16	0.9891
9	B-Random Forest	92.14%	0.01	94.24%	0.01	0.9817
10	B-Random Forest	96.40%	2.70	97.29%	2.73	0.9886
11	B-Naive Bayes	90.44%	0.01	92.73%	0.01	0.9496
12	B-Naive Bayes	91.50%	4.20	93.56%	2.93	0.9527
13	B-SMO RBF Kernel	89.38%	0.01	92.17%	0.01	0.9467
14	B-SMO RBF Kernel	93.88%	3.20	95.40%	3.17	0.9369
15	B-SMO Poly Kernel	90.82%	0.01	93.34%	0.01	0.8816
16	B-SMO Poly Kernel	92.05%	3.50	94.06%	3.07	0.9541
17	J48 Pruned	95.31%	0.01	96.47%	0.00	0.9517
18	J48 Pruned	95.95%	2.77	96.91%	2.16	0.9647
19	J48 Unpruned	95.03%	0.01	96.26%	0.01	0.9523
20	J48 Unpruned	96.12%	2.71	97.04%	2.17	0.9661
21	J48 Reduced Error Pruning	95.10%	0.01	96.31%	0.01	0.9567
22	J48 Reduced Error Pruning	95.93%	2.80	96.89%	2.10	0.9646
23	JRip	94.95%	0.00	96.18%	0.00	0.9534
24	JRip	95.67%	3.13	96.69%	2.34	0.9584
25	Random Forest	92.04%	0.01	94.16%	0.01	0.9813
26	Random Forest	96.26%	2.86	97.19%	2.57	0.9902
27	Naive Bayes	86.26%	0.01	89.76%	0.00	0.9241
28	Naive Bayes	85.50%	6.09	89.17%	2.46	0.9194
29	SMO RBF Kernel	80.15%	0.00	87.07%	0.00	0.7008
30	SMO RBF Kernel	93.83%	3.39	95.36%	3.15	0.9309
31	SMO Poly Kernel	90.90%	0.00	93.41%	0.00	0.8817
32	SMO Poly Kernel	95.45%	3.61	96.58%	2.80	0.9484

Table 19 RWeka results for *God Class* (grey) compared with Arcelli Fontana's (white)

#	Classifier	Accuracy	Std dev.	F measure	Std dev.	AUROC
1	B-J48 Pruned	97.14%	0.00	97.86%	0.00	0.9834
2	B-J48 Pruned	97.02%	2.82	97.75%	2.14	0.9923
3	B-J48 Unpruned	96.68%	0.01	97.51%	0.01	0.9794
4	B-J48 Unpruned	97.02%	2.88	97.75%	2.00	0.9925
5	B-J48 Reduced Error Pruning	97.37%	0.00	98.03%	0.00	0.9885
6	B-J48 Reduced Error Pruning	97.26%	2.64	97.94%	2.18	0.9861
7	B-JRip	96.73%	0.00	97.55%	0.00	0.9879
8	B-JRip	96.90%	3.15	97.67%	2.39	0.9916
9	B-Random Forest	97.55%	0.00	98.17%	0.00	0.9951
10	B-Random Forest	96.95%	2.86	97.70%	2.67	0.9890
11	B-Naive Bayes	94.59%	0.00	95.91%	0.00	0.9757
12	B-Naive Bayes	97.54%	2.65	97.70%	2.72	0.9871
13	B-SMO RBF Kernel	94.92%	0.01	96.21%	0.01	0.9835
14	B-SMO RBF Kernel	94.62%	3.34	95.98%	2.89	0.9838
15	B-SMO Poly Kernel	95.20%	0.00	96.45%	0.00	0.9399
16	B-SMO Poly Kernel	94.33%	3.58	95.75%	2.48	0.9799
17	J48 Pruned	96.61%	0.00	97.46%	0.00	0.9600
18	J48 Pruned	97.31%	2.51	97.98%	1.89	0.9783
19	J48 Unpruned	96.58%	0.00	97.45%	0.00	0.9718
20	J48 Unpruned	97.31%	2.51	97.98%	1.93	0.9783
21	J48 Reduced Error Pruning	97.73%	0.00	98.29%	0.00	0.9761
22	J48 Reduced Error Pruning	97.29%	2.52	97.94%	1.89	0.9742
23	JRip	97.65%	0.00	98.24%	0.00	0.9694
24	JRip	97.12%	2.70	97.81%	2.48	0.9717
25	Random Forest	97.27%	0.01	97.96%	0.00	0.9953
26	Random Forest	97.33%	2.64	97.98%	2.24	0.9927
27	Naive Bayes	94.03%	0.01	95.41%	0.00	0.9820
28	Naive Bayes	97.55%	2.51	98.14%	2.20	0.9916
29	SMO RBF Kernel	89.36%	0.00	92.58%	0.00	0.8435
30	SMO RBF Kernel	95.43%	3.26	96.62%	2.40	0.9427
31	SMO Poly Kernel	95.38%	0.00	96.59%	0.00	0.9410
32	SMO Poly Kernel	95.71%	3.14	96.83%	2.40	0.9459

References

1. Arcelli Fontana F, Mäntylä MV, Zanoni M, Marino A (2016) Comparing and experimenting machine learning techniques for code smell detection. Empir Softw Eng 21(3):1143–1191
2. Fontana FA, Mariani E, Mornioli A, Sormani R, Tonello A (2011) An experience report on using code smells detection tools. In: 2011 IEEE fourth international conference on software testing, verification and validation workshops, pp 450–457.https://doi.org/10.1109/ICSTW.2011.12
3. Fowler M (1999) Refactoring: improving the design of existing code. Addison-Wesley, Boston, MA, USA
4. Grodzicka H, Ziobrowski A, Łakomiak Z, Kawa M, Madeyski L (2019) Appendix to the paper "Code smell prediction employing machine learning meets emerging Java language constructs". http://madeyski.e-informatyka.pl/download/GrodzickaEtAl19.pdf
5. Hall M, Frank E, Holmes G, Pfahringer B, Reutemann P, Witten IH (2009) The weka data mining software: an update. SIGKDD Explor Newsl 11(1):10–18. https://doi.org/10.1145/1656274.1656278
6. Hornik K, Buchta C, Zeileis A (2009) Open-source machine learning: R meets Weka. Comput Stat 24(2):225–232. https://doi.org/10.1007/s00180-008-0119-7
7. Hsu CW, Chang CC, Lin CJ (2003) A practical guide to support vector classification. Technical report, Department of computer science, National Taiwan University. http://www.csie.ntu.edu.tw/~cjlin/papers.html
8. Madeyski L, Kitchenham B (2019) Reproducer: reproduce statistical analyses and meta-analyses. http://madeyski.e-informatyka.pl/reproducible-research/. R package version 0.3.0 (http://CRAN.R-project.org/package=reproducer)
9. Palomba F (2015) Textual analysis for code smell detection. IEEE Int Conf Softw Eng 37(16):769–771
10. Palomba F, Bavota G, Penta MD, Oliveto R, Poshyvanyk D, Lucia AD (2015) Mining version histories for detecting code smells. IEEE Trans Softw Eng 41(5):462–489
11. Palomba F, Nucci DD, Tufano M, Bavota G, Oliveto R, Poshyvanyk D, De Lucia A (2015) Landfill: an open dataset of code smells with public evaluation. In: Proceedings of the 12th working conference on mining software repositories, MSR '15. IEEE Press, Piscataway, NJ, USA, pp 482–485
12. Sharma T (2017) Designite: a customizable tool for smell mining in c# repositories. SATToSE41
13. Tempero E, Anslow C, Dietrich J, Han T, Li, J, Lumpe M, Melton H, Noble J (2010) Qualitas corpus: a curated collection of java code for empirical studies. In: 2010 Asia pacific software engineering conference (APSEC2010), pp 336–345. https://doi.org/10.1109/APSEC.2010.46
14. Witten IH, Frank E (2005) Data mining: practical machine learning tools and techniques, 2nd edn. Morgan Kaufmann, San Francisco

Cloud Cognitive Services Based on Machine Learning Methods in Architecture of Modern Knowledge Management Solutions

Pawel Tadejko

Abstract Cognitive Services are cloud computing services available to help developers build intelligent applications based on Machine Learning (ML) methods with pre-trained models as a service. Machine Learning platforms are one of the fastest growing services of the cloud because ML and Artificial Intelligence (AI) platforms are available through diverse delivery models such as cognitive computing, automated machine learning, model management. Cognitive Computing is delivered as a set of APIs. Due to the nature of the technologies involved in ML ecosystems and Knowledge Hierarchy—Data, Information, Knowledge, Wisdom (DIKW) Pyramid, there is a natural overlap of a technologies and Knowledge Management (KM) processes. The modern architecture of software solutions can be developed with the use of a wide technology stack, including cloud computing technologies and Cognitive Services (CS). We can use a wide range of ML tools at all levels of the DIKW pyramid. In this paper, we propose a new CS based approach to build an architecture of Knowledge Management system. We have analyzed the possibilities of using CS at all levels of the DIKW pyramid. We discussed some of the relevant aspects of Cloud CS and ML in Knowledge Management context and possibilities implementation of Cognitive Services on knowledge processing.

1 Introduction

Machine Learning based content processing in knowledge management approaches often utilize automated feature extraction, detection, and selection techniques. In the business text mining domain, text from documents is often mapped to concepts from an ontology that encodes semantic relationships between concepts. Automatic extraction various information from the massive documents and generation of a profile model are important for a knowledge base. However, the obvious differences

P. Tadejko (✉)
Bialystok University of Technology, Bialystok, Poland
e-mail: p.tadejko@pb.edu.pl

© Springer Nature Switzerland AG 2020
A. Poniszewska-Marańda et al. (eds.), *Data-Centric Business and Applications*,
Lecture Notes on Data Engineering and Communications Technologies 40,
https://doi.org/10.1007/978-3-030-34706-2_9

in structure and content semantics of documents shown by traditional case-base or rule-base systems [1] make it hard to understand critical information.

The need for correlation of distributed semantics model in modern information systems forces to develop more and more intelligent solutions [2]. The connections between ontologies and knowledge modeling become a very promising area [3]. Many of the modern machine learning solutions are already implemented in cloud services provided by the largest providers of cloud platforms like Amazon AWS [4, 5], Microsoft Azure [6, 7], Google Cloud [8], and IBM Watson [9]. Everything indicates that Cloud Cognitive Services will be the basis of business solutions and appears as the future [10] of automated techniques for extraction data, classification of information, and predict or create insights.

Many of the modern machine learning solutions are already implemented in cloud services provided by the largest providers of cloud platforms like Amazon AWS, Microsoft Azure, Google Cloud, and IBM Watson [4, 5, 8, 9, 11, 12]. Everything indicates that Cloud Cognitive Services will be the basis of business solutions and appear as the future [10] of automated techniques for extraction and classification of features.

2 Cloud Cognitive Services Landscape

Cloud Cognitive Computing describes cloud technology platforms that combine cloud computing services, machine learning, reasoning, natural language processing, speech and vision to generate data insights and helps to improve human decision-making [10]. Cloud Cognitive Computing is delivered as a set of APIs that offers a wide range of services to store the data, transform, analyze and visualize.

2.1 Machine Learning Platform as a Service

Most cloud computing services fall into four broad categories: infrastructure as a service (IaaS), platform as a service (PaaS), serverless, and software as a service (SaaS). These are sometimes called the cloud computing stack because they build on top of one another. Machine Learning as a Service (MLaaS) is a general definition of various cloud-based platforms that cover most infrastructure issues such as data preprocessing, model training, deploying chosen model (Fig. 1) and building application with further prediction.

The major cloud computing providers offer nowadays cloud-based AI products. Besides complexity of platforms, we can use high-level APIs. APIs don't require machine learning expertise at all. Currently, the APIs from these four vendors can be broadly divided into three groups: (1) text analysis—recognition, Natural Language Processing, and Natural Language Understanding; (2) image and video analysis—

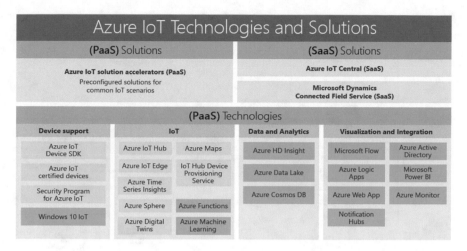

Fig. 1 Example of machine learning platform as a service. microsoft azure IoT framework. Cognitive service cloud computing stack build on top of PaaS/SaaS architecture [13]

recognition, and object detection and identification; (3) other, that includes specific services or domain-based customization of existing service.

The key vendors in the MLaaS space are the large cloud computing providers, including Amazon Web Services (AWS), Microsoft Azure, Google's Cloud, and IBM Watson, among a number of others. These companies offer a huge number of cloud platform and infrastructure microservices and are leading in terms of Artificial Intelligence research.

2.2 Automated Machine Learning

Automated Machine Learning (Fig. 2) is the process of taking training data with a defined target feature, and iterating through combinations of algorithms and feature selections to automatically select the best model based on the training scores. The traditional ML model development process is highly computing-intensive, and time-consuming to compare the results of dozens of models. Automated ML simplifies this process by generating models tuned from the goals and constraints we defined for our experiment [14].

There are some tools and algorithms of machine learning and that we can call ready-to-use technologies, and many of them we can apply directly to knowledge management processes. Using MLaaS enables you to build fast highly customized ML models. Similar to PaaS delivery model MLaaS expects data scientists to bring their own dataset and code that can train a model against custom data.

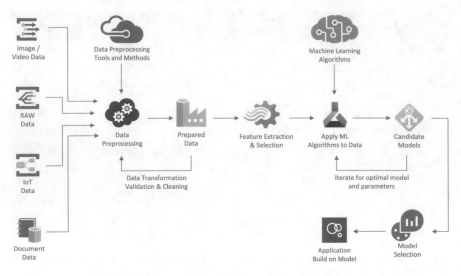

Fig. 2 Automated Machine Learning (AutoML) Reference Architecture (Source: own work based on "Automated Machine Learning Architecture for Predictive Maintenance" [15])

3 Cloud Services Providers and Support for Data and Information Processing for Use in the Knowledge Management Solution

Modern knowledge management systems produce large amounts of data and documents that set new challenges to IT solutions. In certain formats such as document files, images, video, spreadsheets and text appears as unstructured data. Being able to identify entities, people, places, organizations and concepts provides a way for anyone who must make use of the information to understand what it contains. Amazon Machine Learning, Azure Cognitive Services, Google Cloud AI, and IBM Watson are four leading cloud MLaaS services are the four largest players on the cloud scene, and they have all offered many machine learning services, APIs and tools to help us to build solutions powered by AI.

Machine Learning should unlock the knowledge stored in their system to make life decisions easier. After all, "decision making" is a key challenge in Artificial Intelligence and Knowledge Management.

3.1 Knowledge and Insights: Text Processing APIs Comparison

Nowadays, text processing is developing rapidly, and several big companies provide their products which help to build well-fitting solutions with little effort. One of

Table 1 Text processing APIs capabilities comparison: amazon comprehend [5] google cloud natural language [18] microsoft text analytics and microsoft language understanding [6, 19] IBM watson natural language watson natural language understanding [20]

Technique used in information extraction	Amazon comprehend	Google cloud NL	Microsoft text & NLU	IBM watson NLU
Entities extraction[1]	yes	yes	yes	yes
Key phrase extraction	yes	yes	yes	yes
Topics extraction	yes	yes	yes	yes
Topic modeling[2]	2 levels	2 levels	no	5 levels
Intention analysis[3]	yes	yes	yes	yes
Metadata extraction	no	no	no	yes
Relation analysis[4]	no	yes	no	yes
Sentiment analysis[5]	mid	high	low	high
Categories of content	yes	no	no	no
High-level concepts[6]	no	no	no	yes
Entities relations	yes	no	no	yes
Semantic role[7]	no	no	no	yes
Correlate documents[8]	no	no	no	yes

[1] Recognizing names, dates, organizations, etc.
[2] Defining dominant topics by analyzing keywords
[3] Recognizing commands from text (e.g. "run YouTube app" or "turn on the living room lights")
[4] Finding relations between entities
[5] Categorizing how positive, neutral, negative or mixed a text is
[6] Identifying concepts, which can provide more information on related topics than what's in the article; or aren't necessarily directly referenced in the text
[7] Parsing sentences into subject, action, and object form
[8] Building a graph of entities with attributes and use its relationship with other entities to correlate text content across various documents

most popular research area is the law industry [16]. They are very useful even if we implement ML based tools to find information in company's document management system. Researches developed a many concept framework which integrates machine learning, natural language processing and ontology technologies to facilitate knowledge extraction and organization [17]. To understand key pros and cons they have evaluated the usefulness in knowledge management processing (Table 1).

Modern APIs for text processing are very popular and useful. So, if we want to solve specific problem we should use a different APIs. To understand key "pros and cons" they have and when it is better to use one API instead of the other we should compare some features. There are several high-level APIs which may be used to perform these tasks. Among them the most advanced are: Amazon Comprehend, IBM Watson Natural Language Understanding, Microsoft Azure Text analytics API, Google Cloud Natural Language, Microsoft Azure Linguistic Analysis API.

Amazon Comprehend uses natural language processing (NLP) to extract insights about the content of documents without the need of any special preprocessing. It develops insights by recognizing the entities, key phrases, language, sentiments, and other common elements in a document [5]. With Amazon Comprehend you can create new products based on understanding the structure of documents, e.g. search data for mentions of products, scan an entire document repository for key phrases, or determine the topics contained in a set of documents. Amazon even have a special APIs for medical records [11]. To extract insights from clinical documents such as doctor's notes or clinical trial reports, we can use Amazon Comprehend Medical. Amazon Comprehend is NLP set of APIs that, unlike Lex and Transcribe, aim at different text analysis tasks. Currently, Comprehend supports features show in Table 1. This service will help you analyze social media responses, comments, and other big textual data that's not amenable to manual analysis, e.g. the combo of Comprehend and Transcribe will help analyze sentiment in your telephony-driven customer service.

Microsoft Azure suggests high-level APIs, focuses on textual analysis similar to Amazon Comprehend. Language Understanding Intelligent Service (LUIS) is an API that analyzes intentions in text to be recognized as commands (e.g. "run YouTube app" or "turn on the living room lights"). Linguistic Analysis API used for sentence separation, tagging the parts of speech, and dividing texts into labeled phrases. Microsoft Cognitive Services let to build apps with powerful algorithms using just a few lines of code. Microsoft Cognitive Services are a set of APIs, SDKs and services available to developers to make their applications more intelligent, engaging and discoverable. Microsoft Cognitive Services expands on Microsoft's evolving portfolio of machine learning APIs and enables developers to easily add intelligent features such as emotion and video detection; facial, speech and vision recognition; and speech and language understanding into [6, 19].

Google Cloud ML Services set of APIs is a base of Cloud Natural Language API [18] and is almost identical in its core features to Comprehend by Amazon and Language by Microsoft. Google provides AI services on two levels: a machine learning engine (ML Engine) for savvy data scientists [12] and highly automated Google AI Platform [8]. Google doesn't disclose exactly which algorithms were utilized for drawing predictions and didn't allow engineers to customize models. On the other hand, Google is now testing Google Cloud AutoML [21] and suggests using cloud infrastructure with TensorFlow as a machine learning developer's environment. Additionally, Google is testing a number of other popular frameworks like XGBoost, scikit-leran, and Keras [22].

IBM Watson Natural Language Understanding [20] (NLU) is the tool which allows you to train models using your own business data and then classify incoming records. Common use cases are tagging products in e-commerce, fraud detection, categorizing messages, social media feeds, etc. NLU feature set at IBM is extensive. Besides standard information extraction like keyword and entity extraction with syntax analysis, the API suggests a number of interesting capabilities that aren't available from other providers. These include metadata analysis and finding relations between entities. Additionally, IBM suggests a separate environment to train your own models for text analysis using Knowledge Studio. Watson uses a set of transformational technologies which leverage natural language, hypothesis generation and evidence-based learning. Using machine learning, statistical analysis and natural language processing we can find and understand the clues in the questions, and Watson then compare the possible answers, by ranking its confidence in their accuracy, and responded. Watson Engagement Advisor helps organizations know their customers better based [9].

3.2 Knowledge and Insights: Image Processing APIs Comparison

Nowadays, Computer Vision is one of the most widely used fields of machine learning. Building a model for visual recognition is both difficult and time-consuming task. The main task of computer vision is to understand the contents of the image. A large amount of multimedia data (e.g., image and video) is now a significant part of content of the web and knowledge bases. Mining metadata can be very helpful in facilitating multimedia information processing and management.

Fortunately, there are a lot of ready-to-use solutions on the market. They are developed by the various companies like Google, Microsoft, IBM, Amazon, and others. These solutions are namely provided as APIs which you may integrate your apps with. Besides text and speech, Amazon, Microsoft, Google, and IBM provide rather versatile APIs for image and video analysis [4, 7, 7, 23–25]. We are going to make a brief overview of these Cloud APIs capabilities in knowledge management context (Table 2). Some selected and more detailed analysis was described for example in [26]. Preset cloud vision solutions provide more specialized services. One of the example is the face detection API offered by Amazon Rekognition [4] and Microsoft Face API [25]. Interested investigation was provided by [27]. Every vendor has its strengths and weaknesses. But for "common" images all of them have very high success rate in faces detection.

The Google Vision API [23] helps your application to understand what is in the image, classifying the content into the known categories and providing the labels. It is also capable of detecting landmarks e.g. buildings, monuments, natural structures, or logos, performing character recognition that supports a wide variety of languages. The facial detection allows detecting a face with the person's emotion and headwear.

Table 2 Image processing APIs comparison

Technique used in information extraction	Amazon rekognition	Google vision	MS azure comp. vision	IBM visual recognition
Object detection	yes	yes	yes	yes
Scene detection	yes	yes	yes	yes
Face detection	yes	yes	yes	yes
Facial recognition[1]	yes	no	yes	yes
Facial analysis[2]	yes	no	yes	yes
Inappropriate content	yes	yes	yes	no
Celebrity recognition	yes	yes	yes	no
Text recognition	yes	yes	yes	yes
Written text recognition	no	no	yes	no
Brand recognition	no	no	yes	no
Emotions present	yes	no	yes	no
Age & Gender	no	no	yes	no
Food recognition	no	no	no	yes

[1] For detecting faces and finding matching ones
[2] Interesting as it detects smiles, analyzes eyes, and even defines emotional sentiment

The Microsoft Computer Vision API [7] allows classifying the image content by providing a comprehensive list of tags and attempting to build a natural language description of the scene. The Face API is used to detect the faces and retrieve the facial features like emotional state, gender, age, facial hair, smile score and facial landmarks add more complicated features such as emotion and video facial, and vision recognition [6, 19].

Amazon Rekognition [4] can understand what objects and people are in the scene and what is happening. In addition, it can understand the text in the image. One of Rekognition's powers is the ability to detect, recognize and identify people. It is capable of accurate identification of a person in a photo and video using a private dataset of face images. It is also able to analyze the sentiment, age, eye- and headwear presence, facial hair and other features.

IBM Watson Visual Recognition [24] does not have that many models bundled in, but it allows building a custom one. The default one is a general model to understand what objects are in the image, identify the color theme. Another one is for the face detection (not recognition), one is for food detection, and the OCR is in private beta at the moment. The API also allows you to export the model in Core ML (Apple iOS) compatible format.

4   Knowledge Management Processes and Information Technology Systems

The Knowledge Pyramid was developed as a reference model to facilitate systems analysis with respect to facts, data, information and knowledge [28]. Some systems do all their work at one level while other systems might span several levels. Technological systems usually operate in the lower half of the pyramid, but when we use modern tools based on ML methods we can build more advanced tools.

4.1   DIKW Pyramid and Knowledge Processing Level

While looking into the knowledge which pyramid illustrates the relationships of KM to Business Processes, we see the inherent difficulties using IT Systems in order to achieve a competitive advantage in an organization. To discuss the details of the proposed architecture, we must examine the relationship between information systems and the "knowledge pyramid" [28, 29].

IT Systems that operate on the Data Level (DL) may be characterized as those that transform or acquire large amounts of data and documents but have not yet organized this data into a human-useable form. Information-Level (IL) systems can be characterized as having taken the raw data and placed it into structures that can then be searched, accessed, exported and visualized by the users. Only a few years ago, the vast majority of IT systems are currently operating in this field. A subsystem that operates at the Knowledge Level (KL) has organized the information into interrelated forms that are connected directly to an ontology. Present Data Mining models are knowledge-intensive subsystems that enable knowledge extraction and discovery. There are some researches on ontology-based approach for self-describing knowledge discovery [30]. More and more systems today operate at this level, and ML based subsystems allow their users to find knowledge in the information they contain that relates to real world context. The key application of Wisdom Level (WL) subsystem is its ability to model what is truly going on, and to predict or create insights. Unfortunately, there is no information systems that operate at the WL level in generalized domain. A few years ago there were even no solutions in large domain, but with the help of cognitive services we can build some solutions that operate at WL level, and create knowledge and insights.

4.2   Information Systems and Cognitive Services in Model of Knowledge Formation Process

Various IT software tools have been developed over the years to aid Knowledge Management in many market solutions [31] Today, a large amount of advanced tools

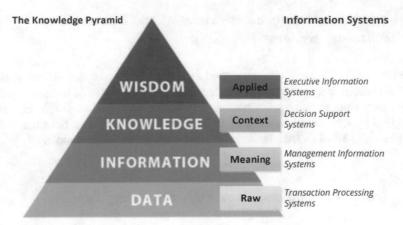

Fig. 3 A DIKW Pyramid based on the different levels of hierarchy in the organization where information systems can support processes of knowledge management. Traditional model of using IT tools in the process of knowledge management [33]

are available that might, at first glance, be helpful for the company, organizations and institutions. However, it appears that not all the tools are equally appreciated by the researches [32].

There are several emerging features (or rather IT tools) in contemporary knowledge management systems. Here's a look at some of the most prominent ones: Content and Document Management, Groupware, Workflow, Lessons Learned Databases, Expert System. All IT tools mentioned above can be used on almost all DIKW Pyramid levels (Fig. 3), respectively "Content and Document Management" on DIKW-data; "Groupware", "Workflow systems" and "Lessons Learned" on DIKW-Information and DIKW-Knowledge; "Expert System" on DIKW-Knowledge and DIKW-Wisdom. Modern "Expert System" are increasingly build on Cognitive Solutions.

Functions of processing level of Information Systems are strictly connected to DIKW pyramid (Fig. 3). Cognitive Services offered by the MLaaS can support more than one level, e.g. natural language processing can be use at the "Information" and "Knowledge" [34]. However, there are still gaps [35] within existing studies on IT-based KM. For example, existing IT tools fail to consider the combination of knowledge with its related objects and the relationship between objects that are affected by knowledge.

We can associate other adjectives with the various layers of the information pyramid, for example if we consider the needs of how information is organized we can divide the pyramid as described in [33]. Nowadays there are two main approaches to process mass volume of data and knowledge to aim knowledge processing objectives: big data analysis and cognitive services. The "big data" approach focuses on the analysis of large amounts of data, their transformation, validation and cleaning. In this case, key mechanisms are associated with data processing processes, but not necessarily with advanced analysis of these data. The engine of such systems is most often based on a rule-based or case-based approach. The "cognitive services" approach focuses on an advanced analysis of data and knowledge, most often with the use of machine learning tools, even without the use of big data. There are far fewer solutions of this class, and actually they are currently at the stage of testing architectures, because cognitive services in the cloud have only been available for several years on the market.

4.3 Information Systems—Knowledge-as-a-Service

These two approaches based on modern SaaS architecture we called Knowledge-as-a-Service. Of course most effective are these that are implemented in cloud computing environment. This is an emerging research trend that constitutes a promising path for organizations. However, only few studies have explored how knowledge management processes are evolve from server's architecture to cloud-computing environments [36]. This research paper attempts to answer this gap by introducing a new approach for the cloud architecture of KMS. But we have room for more types of KMS architecture solutions. The fact that modern knowledge processing systems are based on big data, cloud services or cognitive services does not mean that the concepts of these solutions would already be known. The architecture of KMS solutions has many components that can determine the unique nature of the solution.

The objective of KaaS approach is twofold. First, it aims to develop an architecture of knowledge-based platform using cloud computing paradigms focused on the processing of large amounts of data (big data approach). Second, it aims to design an architecture of intelligent platform using cloud cognitive services in the context of application in processes of all DIKW levels of knowledge management's pyramid (machine learning approach). An overview of challenges related to cloud computing and analysis of cloud mechanisms in the scope of their use for the representation of knowledge, KM processes and knowledge services in the cloud has been extensively investigated in the publication [36].

4.4 Knowledge as a Data Service—Big Data Approach

Big data is perceived as a some kind of digital transformation that move enterprises to the next level into Data-Driven Organizations (DDOs). Knowledge Management Systems use knowledge discovered from big data and merge with traditional organizational knowledge to improve decision-making processes. Many papers propose a service-oriented frameworks for designing a big data-driven KMS to obtain more business value from big data [37].

In order to implement big data paradigm into company, a new generation of knowledge management systems (KMSs) is required to enable insight discovery based on organizational knowledge. In fact, several companies build their competitiveness increasingly on knowledge extracted from huge volumes of data. However, in Big Data context, traditional on-premise data techniques and platforms are not enough [37]. To face the complex big data challenges, only cloud platform can meet the requirements of specific knowledge applications' requirements. However, this raises further problems with KMS in the cloud according to different system layers such as storage, processing, querying, access and management [36, 37].

4.5 *Knowledge as a Cognitive Service—Machine Learning Approach*

The second category of Knowledge as a Service solutions are solutions that use advanced mechanisms and algorithms, starting with semantic networks and ending with the processing of natural language queries and the responses generated by the machine algorithms. Cognitive services are sometimes called MLaaS—Machine Learning as a Services [15, 22]. Using Cognitive Services will help us implement machine learning capabilities in application with no burden of developing complex machine learning algorithms.

An example of this can be the MS Azure QnAMaker [6, 38] categorized under Knowledge AI, service helps build a service with question and answer capabilities from semi-structured or structured contents. In simple words, the QnAMaker service extracts content from a given documents and builds a concrete question and answer database known as a knowledge base. It takes the knowledge base engine to the new level [39]. Unfortunately, use of this type of knowledge base and its maintenance requires a few important steps and new IT solutions architecture. Cloud Cognitive Services is only the tool. To allow to design applications that provide a more advanced features we should design architecture for specific applications.

5 Architecture of IT Knowledge Management System Based on Cognitive Services

The adoption of Machine Learning (more commonly referred as Artificial Intelligence) into Knowledge Management Systems support the human consultants that had been analyzing the data and monitoring the KM processes. Today, cognitive computing, and intelligent data exploration tools, in particular, have huge implications not only for codifying knowledge, and will likely be adopted by more Knowledge Management Systems (Table 3). We know that results in the field of machine learning based solutions are helpful and useful [10, 16, 17].

The possibilities of building dedicated KM systems in the cloud are very wide and we can use cloud services in various configurations. Therefore, there are no universal

Table 3 The knowledge pyramid, Cognitive services & MLaaS APIs as a tools to on each level of pyramid. MLaaS column contains names of text and image processing APIs

Knowledge pyramid	Cognitive services as a KM processing tools	MLaaS APIs
Wisdom	Summarization (sum)	Amazon comprehend
	Paraphrase (par)	Google cloud NL[1]
	Question & answering (Q&A)	IBM watson NLU[2]
	High-level concepts (hlc)	MS azure text analytics
Knowledge	Knowledge discovery (KDD)	Amazon comprehend
	Relation extraction (rex)	Google cloud NL; IBM watson NLU
	Language understanding (NLU)	MS azure text analytics
	Sentiment analysis (sta)	Amazon rekognition; Google vision AI
	Data description (dat)[3]	IBM watson visual recognition
	Language processing (NLP)	Microsoft vision API
Information	Entity recognition (NER)	
	Language processing (NLP)	Amazon comprehend
	Object detection (obj)	Google cloud NL
	Face detection (fac)	IBM watson NLU
	Logo/landmark detection (lol)	MS azure text analytics

[1] Google Cloud Natural Language; [2] IBM Watson Natural Language Understanding
[3] APIs that allow find and recognize people, places, events, and other types of entities mentioned in images and videos content; analyze emotion of specific target or by the document as a whole;

solutions in the field of IT architecture. Unless we mean the patterns of IT system's architecture. A lot depends on the type of data, how they are collected, stored, and the key issue of what a business goal is. Some service-oriented frameworks for designing a system based on big data processing was presented in [37].

However, there is still an important lack of studies examining knowledge services and practices in a cloud computing environment, as well as precise models to facilitate knowledge management in a cloud service approach [36]. Artificial Intelligence techniques, it is necessary to go beyond the shortcomings of the present semi-intelligent conceptual knowledge base proposals.

5.1 Method of Proposed Architecture

The methods are defined as the activities that support the process of organizational knowledge development. A Cloud Cognitive Services KMS operates based on four main activities of the knowledge management process: knowledge capture, organization, transfer and application. We propose covering the processes of DIKW pyramid

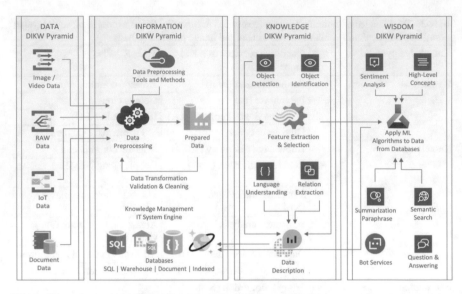

Fig. 4 Uses of cognitive services functions in knowledge management pyramid in relation to layers of the cloud stack: DaaS, IaaS, SaaS, BPaaS (Source: own work)

(Fig. 4) that are similar to functionality of DaaS (Data as a Service), IaaS (Infrastructure as a Service), SaaS (Software as a Service) and BPaaS (Business Process as a Service) layers of the cloud stack.
They are respectively realized as the followings:

- The DaaS and IaaS layer (Data and Information) captures data through the data loading and ingestion components. The data is loaded, processed, cleaned and saved to the data repository. Unfortunately, storing data for the needs of knowledge management systems requires additional post-processing, including semantic indexing, automatic metadata creation, or ontology building.
- The KaaS layer (Knowledge) is able to extract knowledge through the knowledge services. The learning service creates more knowledge for decision-making based on machine learning algorithms. Cognitive Services are a set of APIs that expose the machine learning model to help build intelligent algorithms to hear, speak, recognize, and interpret user input into the applications, websites, and bots.
- The BPaaS layer (Wisdom) delivers the services to applications through combining them with knowledge-intensive business processes. The main purpose of this layer is to use the knowledge services and applications to improve the decision making process. Cognitive Services can be seen as decision-making service to create guideline to make the most effective and strategic business decision.

Classic cloud computing mechanisms are not prepared for specific knowledge processing tasks and require challenging traditional data processing paradigms and taking into account machine learning and cloud technological evolution that dominant current work practices to elaborate adequate knowledge discovery paradigms.

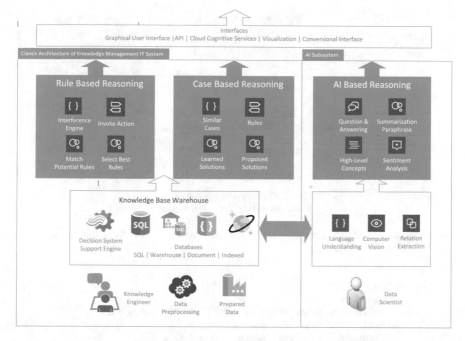

Fig. 5 Cognitive services subsystem in knowledge management IT system architecture (Source: own work)

We need special data storage engines, including ontology building, domain-based indexing mechanisms, semantic data and information retrieval tools, and intelligent text mining tools, query processing and search in knowledge databases using natural language. To try to solve this problem we propose uses of a list of cloud cognitive services (Table 3) and (Fig. 5). Our proposal includes the extension of the classic KMS architecture with the "AI Based Reasoning" module. Current advanced IT systems used in knowledge management contain elements built with the help of machine learning tools, but none of them contain such advanced functions as cloud solutions [10, 31, 39]. The proposed architecture focuses precisely on the possibilities of using cognitive solutions in the cloud and is therefore a new concept at the architecture of KMS class solutions.

5.2 Design Evaluation of Proposed Architecture

The design evaluation is based on the analytical evaluation method to evaluate the utility, quality and efficacy of the proposed architecture. The current concept focuses on the utility of the architecture. Therefore, this evaluation has been carried out in two areas: one for static analysis and one for architecture analysis.

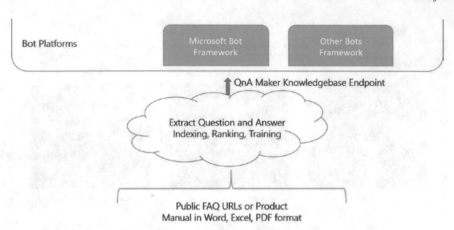

Fig. 6 Architecture of qnamaker service. referred from the microsoft website (Source: Microsoft Azure Documentation [6])

```
{
   "answers": [ {
       "answer":
       "Follow the below steps to embed the QnA Maker service
       Create your knowledge base at
       [https://qnamaker.ai](https://qnamaker.ai) \n*
       Create your Azure service bot :
       [https://docs.botframework.com/en-us/azure-bots/build]
       "score": 26.81,
       "questions": [
          "How do I embed the QnA Maker service in my website?" ] } ]
}
```

Fig. 7 Get answers to a question using a special prepared knowledge base. Referred from the microsoft website (Source: Microsoft Azure Documentation [6])

The first area concerns the static analysis, which examines the artifacts and implementation method two main subsystems: image analysis and NLP. This step has been performed using Microsoft Azure Cognitive Services documentation. We evaluate the way of implementation with Azure Cloud components.

Figure (Fig. 6) presents architecture of "Question and Answer Maker" service based on Natural Language Processing. When we create a knowledge base, every single question and answer is stored in the Azure Search, which is an AI powered cloud search service. QnA Maker automatically extracts question-answer pairs from semi-structured content such as FAQs, product manuals, guidelines, and support documents stored as web pages, PDF, or MS Word files. Example to see how QnA Maker works with question with natural language we see on (Fig. 7).

Figure (Fig. 8) presents architecture of subsystem for analyze images to detect and provide insights about their visual features and characteristics. Azure's Computer Vision service provides developers with access to advanced algorithms that process

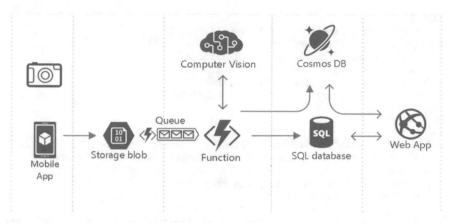

Fig. 8 Architecture of image analyzing based on azure computer vision service. Referred from the microsoft website (Source: Microsoft Azure Documentation [6])

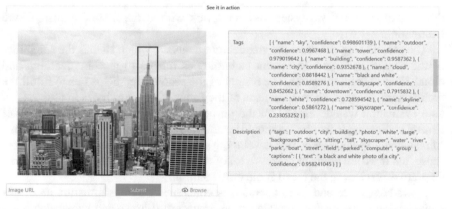

Fig. 9 Microsoft azure. The computer vision feature in action. Referred from the microsoft website (Source: Microsoft Azure Documentation [7])

images. The images processing algorithms can analyze content in several different ways, depending on the visual features you're interested in (Table 2).

Analysis of image (Fig. 9) returns information about visual content found in an image (Fig. 10). For example generate tagging, domain-specific models, and descriptions in four languages to identify content and label it with confidence. Use Object Detection to get location of thousands of objects within an image.

Cognitive services make implementing machine learning easy by consuming APIs. Cognitive services are part of Azure and the Microsoft public cloud. Those cognitive functions as shown in (Table 3) are only available in the cloud environment. Microsoft public cloud (or clouds of other vendors), guarantees a secure, highly available and smooth performance. The implementation of the most advanced features

```
Tags [
{ "name": "sky", "confidence": 0.998601139 },
{ "name": "outdoor", "confidence": 0.9967468 },
{ "name": "tower", "confidence": 0.979019642 },
{ "name": "building", "confidence": 0.9587362 },
{ "name": "cloud", "confidence": 0.8818442 },
{ "name": "skyscraper", "confidence": 0.233053252 } ]

Description
{ "tags":
[ "outdoor", "city", "building", "large", "river", "background", "sitting",
  "tall", "skyscraper", "water", "park", "boat", "street", "computer" ],
  "captions":
[ { "text": "a black and white photo of a city",
    "confidence": 0.958241045 } ] }
```

Fig. 10 Return information about visual content found in an image (a few example tags). Referred from the microsoft website (Source: Microsoft Azure Documentation [7])

(Table 3) of modern KMS systems is not possible without the use of Cognitive Cloud Computing.

The second step concerns the architecture analysis, which studies the suitability of the design artifacts into current technologies related to machine learning and cloud IT system architecture. The results of the second step helped us to fulfill the architecture, to meet the requirements of the future KMS. As shown in (Table 4) ex ante evaluation was developed for the purpose of suitability assessment which of several competing technologies should be adopted if we want to develop Knowledge Management System. Of course, developing small on-premise local KMS still makes sense, but if we are going to process a lot of data and use machine learning methods, Cognitive Services in the cloud is the only solution.

Nowadays, more and more attention is concerned on transformation processes between data, information, knowledge, and wisdom (Fig. 4), and knowledge management system (KMS). Therefore, there are many intelligent processes that convert data into information, information into knowledge, and knowledge into wisdom [40].

Although KM systems can help automate and standardize knowledge management, there are several challenges when implementing a system. Not only technology challenges like security, data accuracy, and changes in technology. A much more pervasive and ongoing challenge, however, is creating a culture of collaboration and knowledge sharing via technology. We propose a new CS based approach to build an architecture of Knowledge Management system. We have analyzed the possibilities of using CS at all levels of the DIKW pyramid. As

While most ML service products have common features, there are plenty that make them unique. Prior to selecting one, you should consider the type of product that you want to build. If we consider as a Cognitive Services in terms of the architecture of IT System for KM solution, at such a high level concept, each of them can be used separately as a MLaaS engine in proposed architecture (Fig. 5). For example, while a cognitive API may be able to identify objects on the image, it can be used to generate

Table 4 Evaluation which of several competing architectural approaches to building a knowledge management system (Source: own work)

Type of KMS Features of KMS	KMS with ML methods engine	Knowledge as a Data service	Knowledge as a Cognitive service
Massive data processing	–	X	X
Machine learning methods	X	X	X
Cloud cognitive services	–	–	X
ML methods as a service	X	–	X
Automated machine learning	–	–	X
Model management	X	–	X
Scalability	–	X	X
Performance	–	X	X
Storage capacity	–	X	X

additional meta data for document description in knowledge base. Assuming you have a large dataset of objects labeled with the make and model, your data science team can rely on ML PaaS to train and deploy a custom model that's tailormade for the business scenario.

6 Conclusion

Knowledge management systems have evolved from a useful tool to optimize some KM processes to an integral intelligent component of KM itself [41]. Today, organizations rely on KM systems to perform many of the functions of knowledge management. Most users have forgotten about basic functions of KM platform such as data storage or workflow [34]. Modern technology systems can also help foster collaboration, intelligent searching, knowledge discovery and generate high-level concepts, among other objectives.

The study in this paper mainly concentrates on the Cognitive Services capabilities and technological aspects of Knowledge Management. We can conclude from the above analysis that Machine Learning as a Services model is mature enough to build advanced Knowledge Management solutions and should be a special subsystem in Knowledge Management IT Systems architecture. There are many different cloud APIs for text analysis and computer vision on the market. In addition, this field is under rapid development. In the article, we made a brief overview of the various providers. At first sight, all of them provide fairly similar capabilities, yet some put an emphasis on some features, or on building custom models like IBM Watson and Microsoft Azure.

The IBM Watson Natural Language Understanding provides a number of "management features" not provided by the others, such as extract meaning—called Macro Understanding—provides a general understanding of the document as a whole. NLU features allow identifying high-level concepts, correlating text content across various document, or even building solution for drafting early phase response documents, e.g. helping legal teams [20].

If we need more specific applications, of course, we can build our own solution, also from scratch. Regardless, the proposed architecture will be still the same. It will differ only in the method of implementing the "AI Based Resoning" subsystem (Fig. 5). In addition, we get the advantages of Cloud Computing. The majority of these APIs provide very high performance, and all of them are subjects for future improvements by the developers.

Acknowledgements This paper was partially supported by grant of Faculty of Computer Science, Bialystok University of Technology, Bialystok, no. S/WI/3/2018 and financed from the funds for science of the Ministry of Science and Higher Education.

References

1. Avdeenko T, Makarova E, i Klavsuts I (2016) Artificial intelligence support of knowledge transformation in knowledge management systems. In: 13th international scientific-technical conference on actual problems of electronics instrument engineering (APEIE), pp 195–201
2. Lupu M (2017) Information retrieval, machine learning, and natural language processing for intellectual property information. World Pat Inf 49, A1–A3, ISSN 0172-2190. https://doi.org/10.1016/j.wpi.2017.06.002
3. Munir K, Anjum MS (2018) The use of ontologies for effective knowledge modelling and information retrieval. Appl Comput Inform 14(2):116–126, ISSN 2210-8327. https://doi.org/10.1016/j.aci.2017.07.003
4. Amazon Web Services, Inc. (2019) Amazon Rekognition intelligent image and video analysis to your applications. http://aws.amazon.com/rekognition/. Cited 10 Apr 2019
5. Amazon Web Services, Inc. (2019) Overview of Amazon comprehend. Amazon comprehend documentation. http://docs.aws.amazon.com/comprehend/index.html. Cited 10 Apr 2019
6. Microsoft Azure (2019) Cognitive services. Language understanding - language understanding into apps, bots, and IoT devices. http://azure.microsoft.com/en-us/services/cognitive-services/language-understanding-intelligent-service/. Cited 10 Apr 2019
7. Microsoft Azure (2019) Microsoft computer vision API. Analyze images and extract the data. http://azure.microsoft.com/en-us/services/cognitive-services/computer-vision/. Cited 10 Apr 2019
8. Google: AI Platforms – Google AI. Machine learning developers, data scientists, and data engineers to take their ML projects (2019). http://cloud.google.com/ai-platform/. Cited 10 Apr 2019
9. IBM: Services Platform with Watson Cognitive Services (2019). http://www.ibm.com/services/technology/platform-with-watson/. Cited 10 Apr 2019
10. Fried J (2019) How knowledge management will change with the advent of machine learning and cognitive search. CIOReview. http://knowledgemanagement.cioreview.com/cxoinsight/how-knowledge-management-will-change-with-the-advent-of-machine-learning-and-cognitive-search-nid-27521-cid-132.html. Cited 10 Apr 2019
11. Amazon Web Services, Inc. (2019) Amazon comprehend medical. Identifies complex medical information. http://aws.amazon.com/comprehend/medical/. Cited 10 Apr 2019

12. Google: Tools – Google AI. Ecosystem by providing tools and open source projects for students and developers (2019). http://ai.google/tools/. Cited 10 Apr 2019
13. Microsoft Azure (2019) Cognitive services. Use AI to solve business problems. http://azure.microsoft.com/en-us/services/cognitive-services/. Cited 10 Apr 2019
14. Microsoft Docs (2019) Automated ML algorithm selection & tuning - azure machine learning service. http://docs.microsoft.com/en-us/azure/machine-learning/service/concept-automated-ml. Cited 10 Apr 2019
15. Deddy L (2019) Automated machine learning architecture for predictive maintenance, pre-senso. AI driven industrial analytics. http://www.presenso.com/blog/Automated-Machine-Learning-Architecture-for-Predictive-Maintenance. Cited 10 Apr 2019
16. Burger S (2019) Machine learning, knowledge management tools changing legal industry, Engineering news, Creamer media. http://www.engineeringnews.co.za/article/machine-learning-knowledge-management-tools-changing-legal-industry-2018-08-29/. Cited 10 Apr 2019
17. Qiao T (2006) Knowledge management using machine learning, natural language processing and ontology. PhD Thesis, Cardiff University
18. Google: Google Cloud. Cloud Natural Language. Derive insights from unstructured text (2019). http://cloud.google.com/natural-language/. Cited 10 Apr 2019
19. Microsoft Azure (2019) Cognitive services. Text analytics - detect sentiment, key phrases, and language. http://azure.microsoft.com/en-us/services/cognitive-services/text-analytics/. Cited 10 Apr 2019
20. IBM: Watson Natural Language Understanding; Natural language processing for advanced text analysis (2019). http://www.ibm.com/cloud/watson-natural-language-understanding/details/. Cited 10 Apr 2019
21. Google: Google Cloud AutoML. Cloud AutoML - Custom Machine Learning Models (2019). https://cloud.google.com/automl/. Cited 10 Apr 2019
22. AlexSoft: Comparing MLaaS - Machine learning as a service: Amazon AWS, MS Azure, Google Cloud AI (2019). http://www.altexsoft.com/blog/datascience/comparing-machine-learning-as-a-service-amazon-microsoft-azure-google-cloud-ai-ibm-watson/. Cited 10 Apr 2019
23. Google: Vision AI - AI & Machine Learning Products (2019). http://cloud.google.com/vision/. Cited 10 Apr 2019
24. IBM: Watson Visual Recognition. Tag, classify and train visual content using machine learning (2019). http://www.ibm.com/watson/services/visual-recognition/. Cited 10 Apr 2019
25. Microsoft Azure (2019) Face API - facial recognition software. http://azure.microsoft.com/en-us/services/cognitive-services/face/. Cited 10 Apr 2019
26. Keenan T (2019) Comparing image recognition APIs, Upwork Global Inc. http://www.upwork.com/hiring/data/comparing-image-recognition-apis/. Cited 10 Apr 2019
27. Torrico DP (2019) Face detection - an overview and comparison of different solutions. http://www.liip.ch/en/blog/face-detection-an-overview-and-comparison-of-different-solutions-part1/. Cited 10 Apr 2019
28. Marjanovic O, Freeze R (2012) Knowledge-Intensive business process: deriving a sustainable competitive advantage through business process management and knowledge management integration. Know Process Mgmt 19:180–188. https://doi.org/10.1002/kpm.1397
29. Pathak AR, Pandey M, Rautaray S (2018) Construing the big data based on taxonomy, analytics and approaches. Iran J Comput Sci https://doi.org/10.1007/s42044-018-0024-3
30. Konys A (2018) Knowledge systematization for ontology learning methods. Procedia Comput Sci 126, 2194–2207, ISSN 1877-0509. https://doi.org/10.1016/j.procs.2018.07.229
31. Sawant A (2018) Market research future. Knowledge management software market 2018, regional study and industry growth by forecast to 2023. http://www.marketwatch.com/press-release/knowledge-management-software-market-2018-global-size-competitors-strategy-regional-study-and-industry-growth-by-forecast-to-2023/. Cited 10 Apr 2019
32. Choenni S, Harkema S, Bakker R (2005) Learning and interaction via ICT tools for the benefit of knowledge management. In: Baets W (eds) Knowledge management and management learning. Integrated series in information systems, vol 9. Springer, Berlin. https://doi.org/10.1007/0-387-25846-9_7

33. Kimble C (2018) Types of information system and the classic pyramid model. In: World med mba program - information systems and strategy course (2018) Available via personal website. http://www.chris-kimble.com/Courses/. Cited 10 Apr 2019

34. Chugh M, Chugh N, Punia A, Agarwal DK (2013) The role of information technology in knowledge management. In: Proceedings of the conference on advances in communication and control systems-2013

35. Crane L (2015) Knowledge management's theories. In: Crane L (eds) Knowledge and discourse matters, Chapter: key issues and debates, pp 37–50. https://doi.org/10.1002/9781119079316.ch4

36. Depeige A, Doyencourt D (2015) Actionable knowledge as a service (AKAAS): leveraging big data analytics in cloud computing environments. J Big Data 2(1):12. https://doi.org/10.1186/s40537-015-0023-2

37. Le Dinh T, Phan TC, Bui T, Vu MC (2016) A service-oriented framework for big data-driven knowledge management systems. In: Borangiu T, Dragoicea M, Novoa H (eds) Exploring services science. IESS 2016. Lecture notes in business information processing, vol 247. Springer, Berlin

38. Shaikh K (2019) Eagle-eye view of azure cognitive services. In: Developing bots with QnA maker service. Apress, Berkeley, CA

39. Zarri GP (2018) High-level knowledge representation and reasoning in a cognitive IoT/WoT context. In: Sangaiah A, Thangavelu A, Meenakshi Sundaram V (eds) Cognitive computing for big data systems over IoT. Lecture notes on data engineering and communications technologies, vol 14. Springer, Berlin

40. Wognin R, Henri F, Marino O (2012) Data, information, knowledge, wisdom: a revised model for agents-based knowledge management systems. In: Moller L, Huett J (eds) The next generation of distance education. Springer, Berlin. https://doi.org/10.1007/978-1-4614-1785-9-12

41. Chalmeta R, Grangel R (2008) Methodology for the implementation of knowledge management systems. J Am Soc Inf Sci Technol 59(5):742–755. https://doi.org/10.1002/asi.20785

Anti-Cheat Tool for Detecting Unauthorized User Interference in the Unity Engine Using Blockchain

Michal Kedziora, Aleksander Gorka, Aleksander Marianski
and Ireneusz Jozwiak

Abstract The purpose of this paper was to analyse problem of cheating in online games and to design a comprehensive tool for the Unity engine that detects and protects the applications against unauthorized interference by the user. The basic functionality of the tool is detecting and blocking unauthorized interference in the device's memory, detecting the modification of the speed of time flow in the game and detecting time changes in the operating system. The research also examined the current potential of blockchain technology as a secure database for the game. Proposed algorithms where implemented and tested for usability and speed of operation.

1 Introduction

Along with the popularization of smartphones, a number of mobile games with multiplayer elements such as rankings, tournaments and competitions, in which other users participate is rising rapidly. In games of this type, especially those created by small developer studios, ability of users cheating is simplified, because most of the operations are performed on the device on the client side, and only results are sent to the server. This resulted in the creation of many tools enabling cheating in a very simple way that can be used by any person. Most of the mobile games even those one-person earns by means of micropayments, so any fraud in such a game causes financial losses for developers. Instead of buying a given item or improvement, the player will get it using cheat. Losses in multiplayer games will be even greater because the player who in an illegal way it will be in a high place with a ranking with a disproportionately high score may discourage honestly playing players for further play, and thus to spend money in it.

The best example of the size of the computer games industry is the fact that the game World of Warcraft, which premiered in 2004 to date has earned $ 9.23 billion,

M. Kedziora (✉) · A. Gorka · A. Marianski · I. Jozwiak
Faculty of Computer Science and Management, Wroclaw University
of Science and Technology, Wroclaw, Poland
e-mail: michal.kedziora@pwr.edu.pl

© Springer Nature Switzerland AG 2020
A. Poniszewska-Marańda et al. (eds.), *Data-Centric Business and Applications*,
Lecture Notes on Data Engineering and Communications Technologies 40,
https://doi.org/10.1007/978-3-030-34706-2_10

and the game GTA 5, which premiered in 2015, has earned over $ 6 billion and has been distributed in 90 million copies. None of the films ever made even reached half that revenue. A significant portion of GTA 5's revenues was generated through micropayments in multiplayer. However, many players used cheats to facilitate their gameplay and giving an advantage over honest players. Any cheat in a multiplayer game generates a loss. One of the creators of cheats for GTA 5 was convicted by a court in the USA because the software he created generated losses for the company Take-Two Interactive at half a million dollars.

An example of a universal cheating tool, which additionally has versions for Android and Windows is Cheat Engine. Its main functionalities are searching and editing variables in the device's memory as well as accelerating the flow of time in the game. It is able to search for variables of all basic types (int, float, string, double, bytes, etc.). Cheat Engine is a free program released under the free software license [1]. The next programs that have the same functions as Cheat Engine are SB Game Hacker APK, GameGuardian and Game Killer. The main function of all these programs is the manipulation of the device's memory, specifically the search for appropriate variables from the game and their editing to the values given by the user. Apart from interfering with the memory of the device, these programs are also able to change the speed of time in the game [2]. Another type of cheating tools are programs that allow users to bypass the payment requirements in the application. Additionally with the help of interfering tools in the payment system you can make unlimited purchases. Programs offering these functionalities are Creehac and Freedom APK [3]. The basic function of all these programs is to search and change the data stored in the device's memory.

2 Classification of Cheating Methods

Cheating in a computer game, we define as breaking the rules set by the authors of the game. The purpose of cheats is to quickly win the prize and gain an advantage over the opponent, whether it is an artificial intelligence controlled by a computer or the other player [4]. User can cheat in many different ways. One of them is using codes implemented by the authors to accelerate and facilitate testing during the production of the game. This technique is used in single-player games, therefore it does not generate financial losses for game developers and does not adversely affect its perception, and in some cases can add variety to the game. Another way of cheating is to use external software that affects the elements of game mechanics. This method is mostly used and mainly occurs in multiplayer games. It is negatively treated by the community and can significantly affect the developers finances. In addition, the player using such "help" can be banned, which will result in blocking his account and preventing the game. The ban can be temporary or permanent, which will involve the loss of the game, and therefore with the loss of money that has been invested in it. The last way to cheat in computer games that has not been included in this research is to use the bugs of the game itself. This method works well in both single and multiplayer games. However, it is much less frequently used than external software because finding an error, the use of which can bring any benefits in the game is not easy. Moreover, when such an error is found and made available to a wider group of people, the creators usually quickly release patch to fix software.

2.1 Exploiting Misplaced Trust

An example of this is to give the player too much data that can be used to gain advantage over the opponent. If the player has access to the games logic code or its configuration files, he can easily change them to his advantage. Examples of cheats using this technique are: maphacks (show places where other players or items on the map are located) and wallhacking (allow players to see through the walls, and in some cases even going through walls). Both involve the interception of data that is sent from the server to the client, their appropriate modification and sending them back to the server [5]. These types of cheats also include various types of related frauds with interference in the devices memory. They most often modify the attributes of the player, items in the game world or even global parameters of the game world such as the speed of time flow or the size of the gravitational force. They are even used in games where there are no opponents. Then the most common purpose of the cheater is to speed up the overall pace of the game, which leads to faster gaining appropriate benefits in the game [6].

2.2 Abusing the Game Procedure

This deception is described as a situation in which the player abuses the way the game works and how the player is punished and rewarded. The use of such abuses causes the benefits to be tilted, which should be random to the fraudster's side. An example here is the interruption of the connection from the game in the event of a loss. This will result in not being recognized in the player's ranking, which will make the ratio of winnings to losers false [5, 7].

2.3 Exploiting Machine Intelligence

This deception involves the use of artificial intelligence techniques to gain an advantage over the opponent. An example here can be the use of a game simulation chess with artificial intelligence, while playing with a real opponent. It is enough to copy the opponent's moves to the game board against the advanced artificial intelligence algorithm, and then mirror the movements of the "computer" on the board on which we play with the opponent. Another popular example of cheat using artificial intelligence are aimbots (automatic targeting of the opponent, most often used in FPS games such as first person shooter) or triggerbots causing automatic shooting to the opponent when the player is aiming at him, provides worse results than aimbot, but detection is very difficult [5, 8].

2.4 Modifying Client Infrastructure

The way cheats work using the client's infrastructure is very similar to those described above based on poorly placed trust. However, it does not base actions on modifying the game client and does not use the data received from the server to work. In this case, the device or its drivers are attacked. An example here can be wallhack, which can be implemented by modifying the graphics card drivers, which will change the transparency of the walls [6].

2.5 Exploiting the Lack of Secrecy

This type of fraud can be done in a game where there is no system that cares about the secrecy and integrity of data sent to the server. In this case, the communication packets can be intercepted, changed and sent to a server which, due to the lack of an integrity checking system, will allow entering incorrect data into the database. In addition, if important data is stored in the device's memory in an unencrypted form, trust will also be incorrectly placed, which may lead to even easier interception and modification of data by a fraudster [5]. However, this type of fraud does not only apply to data processed by the game but also to its source code. There are suitable programs that are able to reverse engineer the software. After its completion, the fraudster has access to the source code of the game client, which he can freely change. Then he can run the modified game. This method gives unlimited possibilities and can be used to create virtually any cheat [6].

2.6 Exploiting Lack of Authentication

In a game where there is no proper authentication mechanism, users can collect information about other players' accounts by using a fake server to which they will log in. In addition, after connecting to such a server, the scammers can modify their data, and other players' data and in this way unblock things that they should have access to after making a micropayment. Another case is insufficient server security, which may not stop the cheater from interfering with its functions and gaining access to the database [7, 9].

2.7 Timing Cheating

This method involves delaying the sending of packets by the fraudster, but it does not affect the time of receiving packets. The cheater and other players on his device are not delayed, but other players on their own devices notice the cheater with a considerable delay. Thus, a cheat player has much more time to react, thanks to which he gains the advantage [5].

2.8 Human Factor and Other Methods

All the cheats described above belong to the group of technical frauds and they are characterized by the fact that whether a given fraudster manages to violate the system depends on the strength of security implemented by the authors of the software on both the game and server side. In contrast, frauds in this group are caused by a human factor and the security implemented by the developers have little impact on them. The first type of fraud is account theft, which is caused by carelessness or naivety of the players. Fraudsters use social engineering methods, e.g. phishing, password harvesting fishing, or software to get account data. However, it must be somehow delivered to the player's computer of the game, most often via files downloaded from the Internet for example: guides, add-ons or cheats created by cheaters. Such software includes implemented e.g. keylogger, which registers subsequent keys pressed by the player, and then sends them to the cheater's computer. Collusion fraud is another type. It occurs when a group of two or more dishonest players, e.g. from opposing teams, starts exchanging information that should not be known to the other party. The last type is fraud related to internal abuse. It can be committed by people responsible for the functioning of the game, e.g. administrators or game masters. It occurs when people with more rights than a regular player make it easier to play for selected units, e.g. their friends [7, 9].

2.9 Related Work

A number of papers [10–14] was introduced with aim to research issue of cheating in online games. Detailed systematic classification of cheating in online games was presented by Jeff Yan [9]. Also several tools are already implemented. The most popular tool created for the Unity engine to prevent cheating is the Anti-Cheat Toolkit. It offers functionalities such as protection of variables in memory, provides recognition of the basic types of wallhack, detects flow speed and time changes, protects and extends the built-in system to write to a file named PlayerPrefs [15]. Another tool available to the Unity engine to counteract cheating is PlayerPrefs Elite. The functionalities aimed at improving the security of game data offered by this program are: detection of unauthorized modification of data stored in the PlayerPrefs system, provision of a secure key manager, and encryption and decryption of variables stored in PlayerPrefs. In addition, the tool extends the PlayerPrefs system provided by Unity libraries with the ability to store arrays and vectors. The next tool similar to previous programs is Anti-Cheat—Protected Variables. Its basic application is protection of fields and properties from memory manipulation, protection of data stored in the PlayerPrefs system and detection of speedhacks. Another existing solution that provides functionality different from the previously described tools is Obfuscator Pro. However, it does not give direct protection against cheaters by detecting them or blocking interference in the program. Its main functionality is to obfuscate the program code by adding a random code or changing the names of classes, variables, functions etc. to unreadable ones. Thanks to this, it provides protection against fraudsters' intrusion into the program code. In contrast, the code itself does not provide virtually no protection against basic frauds and it is necessary to extend it with tools that provide basic functions to protect variables or detect other frauds. In summary, no tool includes both anti-fraud and anti-malware functions. In addition, none of the tools found offer an additional form of data protection, e.g. by storing them on the server, and non of presented solutions uses blockchain technology for this purpose [16–18].

Scams based on artificial intelligence can be detected in three ways. In the first of these, you need other players who, if they observe someone suspicious behavior, can report it through the appropriate system. When a suspicious player is reported by a sufficient number of people, he is checked by the game administration or other most experienced players. People who check potential cheaters have access to their statistics and to video recordings of the game. After looking at fragments of games and statistics, such people decide whether or not a person has illegal support software. The difficulty of detecting cheats based on artificial intelligence depends more on the cheater himself. If he sets the aimbot to maximum effectiveness. So he will automatically aim at unnaturally fast pace to every player who is within the range of the shot, he will be unmasked and banned very quickly. However, a cheater who will use a triggerbot or even aimbot, but in a thoughtful way. So he will set the program so that automatic aiming is activated only when another player is very close to the crosshair, even a person watching the video from the game may not notice that the cheater is using illegal software. The described system can be found e.g. in the game Counter-Strike: Global Offensive where it operates under the name Overwatch.

The effectiveness of the system described above is estimated at 15–30%. To achieve much greater efficiency at the level of 80– 95%, it is necessary to use methods of artificial intelligence, deep learning or dynamic Bayesian networks. These methods, having the appropriate test sets, are able to learn "normal" gameplay, and thus will able to recognize deviations from it, which may be caused, among others, by the use of cheats or the use of game errors. However, a large amount of data is needed for the machine learning system to function properly. The Valve company responsible for Counter-Strike: Global Offensive collected all data from the previously described Overwatch system. Thanks to this, she was able to implement the VACnet system (Valve AntiCheat), which uses deep learning methods to recognize cheaters. However, for the correct operation of the system based on artificial intelligence or machine learning methods for a large number of players in real time, you need a lot of computing power. For example, Valve AntiCheat needs up to 1,700 processors to work properly [8, 19].

The first and basic barrier that should be used to prevent a fraudster from reading data should be to encrypt it. The data stored in the device's memory should be encrypted and transferred between the client and server. In addition to the data, the scammer may have access to the source code. However, obfuscation transformation will make reading the game client code much more difficult. This technique involves transforming the program to preserve its semantics and at the same time significantly impede understanding. Code obfuscation can be done in many different ways, here are a few: formatting, changing variable and function names, deleting comments, splitting, joining, changing the order of arrays, cloning methods, changing inheritance relationships. The problem with the method of obfuscating the code is that, although it will make it harder to understand the code, the determined attacker will eventually find out how it works. The last and most effective way to secure important game data is to perform all significant operations on that data on the server. Even if the fraudster gains access and changes the data in the device's memory, the server will still perform the operation on its own data and then update it with the fraudster. The main disadvantage of this solution is that the game must have a server with which each client has a permanent connection during the game [6].

The game design itself has a large impact on the occurrence of fraud based on the abuse of procedures and the use of errors. You can prevent such cheats by implementing functionalities into the game that will make these cheats unprofitable for the player. An example of such functionality may be the introduction of a penalty system for players leaving the game during its duration. However, with these types of scams, it is more important and more difficult to detect them, because before developers can fix something, they need to know what is working incorrectly. Detection of this type of error occurs by collecting various types of data about players, and then their appropriate association by analysis carried out by people or programs using methods of artificial intelligence. In this way, for example, you can detect the connection between players who have a higher win/lose ratio than players who leave the game more often [6].

3 Proposed Anti-Cheat Solution

The solution has been designed for games containing multiplayer elements such as rankings, tournaments and trade between players, created mainly for mobile devices. At the moment there is no generally available aimbot or triggerbot operating on a mobile device. For this reason, the project has omitted protection against fraud by using artificial intelligence algorithms for its operation. The most common feature provided by almost all cheating tools available on the Android system is to search for cached data and freely change them. Therefore, the main functionality of the implemented software will be to detect attempts to interfere with this data and to prevent this interference. This protection will be implemented in two versions. An online version using blockchain and an offline version that uses the cryptographic hash function to check the integrity and authenticity of the data. Another feature commonly available in various types of cheats is speedhack. It allows the player to increase or decrease the speed of time in the game, which can significantly affect the speed of the player's progress. For this reason, one of the additional functionalities of the designed anticheat algorithm will be the detection of such interference in the speed of time flow. Often, games designed for mobile systems have a system of awarding players daily login rewards or depend on many factors since system time for example, the amount of raw materials the player receives or the time it takes to expand buildings. Therefore, very frequent cheating in games is interference in the system time of the device. It does not require any additional software and is very easy to carry out, and in unprotected games can bring great benefits to cheating players. In the designed tool, the appropriate function will be implemented that communicates with the server from which time will be collected.

3.1   Detection of Interference in the Device Memory

Tampering with the device's memory can be hindered and detected in several different ways. Effectiveness of method depends mainly on whether the game processes data on the client or server side. In games that do not have any multi-player elements the implementation of additional systems could transfer only negative effects, e.g. in the form of a decrease in performance. Another situation occurs when the game has multiplayer elements, but the whole logic is performed on the client's device, and only players results are sent to the server. In this case, implementation of solutions detecting cheating is necessary. However, due to the fact that data operations are performed on the client's device, the implemented methods will never give 100% resistance. However, they may make it more difficult to intervene in the data so that it will be unprofitable, and for most people who do not have adequate knowledge and skills will make it impossible. Locally, data can be secured in several different ways. One of them is to duplicate data and store it in several versions, and then compare whether these versions differ from each other. Different versions mean that one of them has been changed unauthorized. Another way is to encrypt data stored in memory and decrypt them when read and changed. For encryption we can use simple algorithms, but some cheat programs are able to find and change even such encrypted data. The last type of security can only be used in multiplayer games that require a constant connection to the Internet. It consists in storing all data and performing all operations on the server, while the game client only provides a graphical interface. In the our proposed solution, instead of a standard server, a blockchain was used to ensure greater security and greater data integrity.

The first step is the installation and configuration of the Loom environment. Launched locally, DAppChain acts as the side chain of the Ethereum Blockchain network. The Loom SDK is a basic component of the Loom network. The Loom network is a platform for building scalable side chains in Ethereum, with particular emphasis on social games and applications. The Loom SDK helps developers create their own blockchain without having to understand the details of its infrastructure. The Loom SDK generates DAppChain, which uses Ethernum as the base layer. Launching DApp as an Ethernum side chain (sidechain) enables the use of an alternative set of rules such as DPoS to achieve high scalability while providing security guarantees at the Ethernet level. Delegate Proof of Stake (DPoS) is a consensus algorithm that relies on a reputation system and continuous voting through which a team of trusted entities (delegates) is selected. Delegates can create and add blocks to the chain and include specific transactions in the blocks. All their activities are publicly available, and real-time digital democracy will promptly dismiss an inefficient or malicious delegate. All services offered by the Loom network use LOOM tokens. Tokens can be used as currency in Loom SDK-based applications and games. In addition, they are needed by developers whose application uses the Loom network to pay for licenses and side chain usage bandwidth measured in transactions per second. Loom does not use player devices to process transactions. Therefore, each transaction is indirectly processed by the Ethernum network which is connected with the generation of costs charged to the developers whose game uses the side chains of this network. Next, the Loom SDK package was imported to the Unity environment and the appropriate functions connecting and communicating with DAppChain were implemented.

```
async void Start()
{
  if (privateKey == null)
  {
    privateKey = CryptoUtils.GeneratePrivateKey();
    publicKey = CryptoUtils.PublicKeyFromPrivateKey(privateKey);
  }
    var contractEvm = await GetContractEVM(privateKey, publicKey);
    await StaticCallContractEVM(contractEvm);
    contractEvm.EventReceived += ContractEventReceived;
    await CallContractEVM(contractEvm);
}
```

The Start() function is executed when the application is started. Its task is to generate the private and public key needed for communication with blockchain. These keys must be unique to each user from the application. Next, an intelligent contract instance is created based on the keys. When you connect to DAppChain via contracts, you can send and save data in it. The status of a smart contract can be changed by signing a transaction and sending it to blockchain, which verifies its correctness. Most of these functions are performed using the Contract.CallAsync() method provided by the Loom SDK. In the body above the method, the function of an intelligent contract is defined. During the implementation, the GetMsg() and SetMsg() functions were used. Their task is to collect the value stored under a given key in sequence and change the value assigned to a given key in DAppChain.

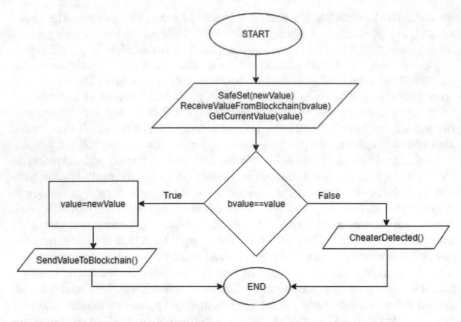

Fig. 1 Algorithm for detecting device memory interference using blockchain

```
private async Task CallGetContract(Contract contract)
{
  var result = await contract.CallAsync<MapEntry>(''GetMsg'', new MapEntry
    { Key = this.key });
}
private async Task CallSetContract(Contract contract)
{
 await contract.CallAsync(''SetMsg'', new MapEntry
   {
    Key = this.key,
    Value = this.value.ToString();
   });
}
public async void SafeSet(int value)
{
  await CallGetContract(contract);
   if (this.blockchainValue != this.value)
    {
     Debug.LogError(''CHEATER DETECTED!'');
     return;
    }
   this.value = value;
   await CallSetContract(this.contract);
}
```

To protect against fraudsters interfering with the device's memory, a special algorithm has been implemented, the scheme of which is visible in Fig. 1. Every time the program intervenes in a variable, this algorithm is executed. The algorithm implementation function retrieves values from the blockchain and compares it with the current value stored in the memory. If they are the same, the variable is changed and the contract executed, which updates the value in the blockchain. The problem in this case is that variables stored in DAppChain can not be updated too often, because this will cause too much load on the blockchain.

3.2 Data Integrity Protection

Earlier we described two methods of data protection against unauthorized user interference. The first was to keep a copy of the data, but if the copy of the data is kept in the same form as the original, with the help of the appropriate software user can easily find and change it with the original. The fraud will not be detected then. The second method was data encryption. But the use of simple encryption algorithms can be broken by a suitably advanced cheat, and the use of advanced encryption methods can affect performance. For these reasons, a combination of these two solutions has been implemented as in Fig. 2. Individual variables will be copied and a hash function will be called for them, and then, when performing each operation on a particular variable, it will be compared with the original. If a difference is detected, it means that the data has been changed in a different way than with the authorized operations.

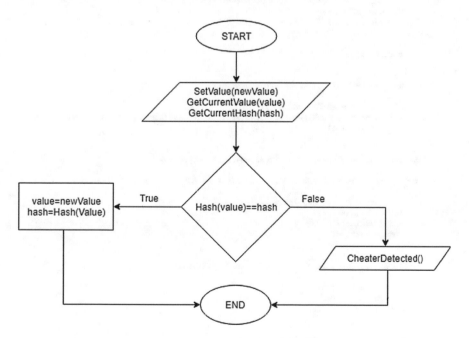

Fig. 2 Algorithm for detecting memory tempering using cryptographic hash functions

For each of the basic types, the overriding class has been implemented e.g., the int has been replaced by the HashInt class.

```csharp
public class HashInt : IEquatable<HashInt>
{
    public int Value;
    private byte[] hash;

    private HashInt()
    {
        Hash(this);
    }

    private HashInt(int value)
    {
        this.Value = value;
        Hash(this);
    }

    private static void Hash(HashInt hashInt)
    {
        var sha = new SHA256Managed();
        hashInt.hash = sha.ComputeHash(Encoding.UTF8.GetBytes(hashInt.Value.ToString()));
    }

    private static bool IsCheated(HashInt hashInt)
    {
        if (!Compare(hashInt))
        {
            Debug.LogError(''Cheater detected!'');
            hashInt.Value = 0;
            return true;
        }

        return false;

    }
    private static bool Compare(HashInt hashInt)
    {
        var sha = new SHA256Managed();
        return hashInt.hash.SequenceEqual(sha.ComputeHash(Encoding.UTF8.GetBytes(hashInt.
    Value.ToString()))));
    }

    #region operators, overrides, interface implementations

    public static implicit operator HashInt(int value)
    {
        var hashInt = new HashInt(value);
        IsCheated(hashInt);
        Hash(hashInt);
        return hashInt;
    }

    public static implicit operator int(HashInt hashInt)
    {
        IsCheated(hashInt);
        Hash(hashInt);
        return hashInt.Value;
    }

    public bool Equals(HashInt other)
    {
        return other.Value.Equals(Value);
    }
```

```
public override string ToString()
{
    return Value.ToString();
}

#endregion
}
```

Each class has two fields, one with the value in the normal form, and the other with the value in the form of a hash. As a hash algorithm, the 256-bit SHA-2 version was used. The key role in the classes is in the functions overwriting all basic operators, because before each of them is called, and therefore when each attempt to edit the variable is checked the compatibility of the value field with hash. If the fields do not match, then the variable has been edited unauthorized and it is reset. In the event that the game connects to the server in this place, a function can be placed that sends information that the player has tried to commit fraud. This could result, for example, in placing the player on the black list and not taking it into account in the rankings.

3.3   Timing Cheating

The speed of playing time can be changed in two ways. The first is to find and change the field stored in memory, responsible for the time scale. Second way to change the speed of the game is more complex. To understand it, you should know that each program communicates with the operating system kernel using predefined functions called system calls. Each operating system has a different set of calls, but often performs similar tasks such as allocating memory, reading and writing files or handling processes. One of the key information processed by the system is knowledge of time. If a computer game is to render 60 frames per second, the rendering function is called average every 16.6 ms. In Windows, the GetTickCount() function returns the number of milliseconds that have passed since the appearance of the system. Thus, the second way to Speed Hacking works by injecting the code into a running process and taking control over the timing function. The purpose of this is to return accelerated or slowed ticks so that you modify the speed of the program. However, the use of this method causes a "crossover" between the time from the start of the game measured by the game environment and the one measured by the operating system. This knowledge has been used to implement a function that detects cheating rate swings [1]. At the moment of starting the program, the start time taken from the system and the game environment is saved. Data is saved and also compared in the form of a number of ticks, in order to achieve the highest possible accuracy. The IsSpeedhackDetected() function is called in the Update() function, which is performed in every game frame (Fig. 3).

Fig. 3 Algorithm for detecting Timing cheating

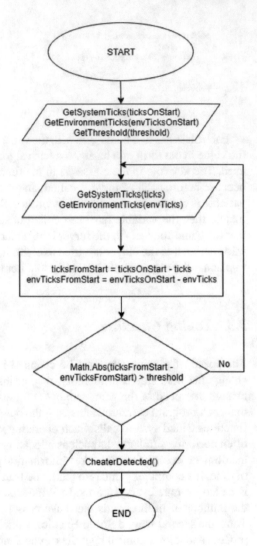

```
private bool IsSpeedhackDetected()
{
    var ticks = System.DateTime.UtcNow.Ticks;

    if (ticks - previousTicks < INTERVAL_TICKS) {return false;}
    previousTicks = ticks;
    var ticksFromStart = ticks - ticksOnStart;

var vulnerableTicks = System.Environment.TickCount *
System.TimeSpan.TicksPerMillisecond;

var ticksCheated = Mathf.Abs((vulnerableTicks
- vulnerableTicksOnStart) - (ticksFromStart)) > THRESHOLD;
```

```
if (ticksCheated)
{
    ResetTicks();
    return true;
}
    return false;
```

The Update() method is the basic function in the Unity engine. Thus, the program in each frame checks if the system time taken using the System.DateTime.UtcNow. Ticks property differs from the time of the currently used environment that is accessed by using the System.Environment.TickCount property. If the difference between these values rises above the set threshold, the change in the speed of time will be detected.

3.4 Modifying Client Time Settings

Time and date change detection refers to games that reward a player for coming into play at a given time. Thus, the method described earlier will no longer work because the player accrues bonuses when he is not present in the game and he is rewarded when entering the game. Thus, in this case, the game very often uses the system date and time, which can easily be freely changed in the system options. This can be prevented by downloading the time from the server and comparing it with the system time of the device. The TimeCheatingDetector class has been implemented, which provides a static Start Detection function. This function, at some fixed default for one minute, takes time from the server and checks if it differs from the system time by the amount defined as the threshold and set by default to 5 min (Fig. 4).

4 Practical Tests

Testing of the implemented functionalities have been carried out with the help of Cheat Engine software. For the function of hashing the variable copy, an attempt was made to find the appropriate variable in memory and to edit it later. Undoubtedly, the results obtained were very interesting. While trying to find a variable, it was not found by the program. The reason for this was the function overwriting the operators, because after using any of them a new object was created that overwrote the old one. Thus, each change in value was also associated with the change of the address of the variable. This prevented the Cheat Engine program from narrowing the number of addresses, which resulted in the lack of finding the right one. To test the effectiveness, a special test function was implemented that changed the value without creating a new object. After this change, the program found the value after three scans, but unauthorized change was detected by the implemented classes each time. The results of finding a given field by the Cheat Engine for the fixed and

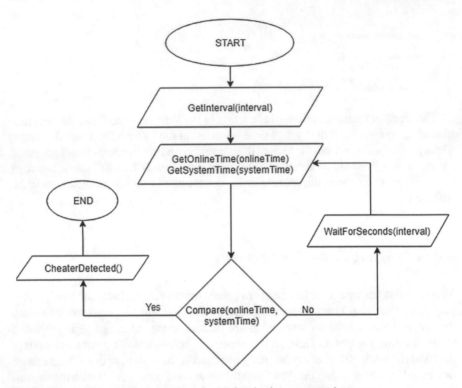

Fig. 4 The algorithm for detecting time change fraud using an external server

variable address of the variable are presented in Table 1. The tests have shown that the solution works in the intended manner, and moreover, it has not one but two barriers to interfering with memory. The first one is dynamically changing value indicators, making it difficult for programs to find the right address, and the other hash copies of variables that ensure the integrity and accuracy of the data. Next the test was carried out for the function detecting the change of the game time speed. The Cheat Engine program speedhack was used, and then the speed has been set several times to both values less than 1 and bigger. At the detection threshold of 500 ms, the detection time was practically instantaneous. However, the change in detection threshold for 5 s detection time is so long that it creates the theoretical possibility of cheating. The player or program can jump from the acceleration value 2–0.5 every 2.5 s and in this way after every 5 s the difference between the time measured by the environment and the time returned by the system will be 0. Implementation of such a system for cheating would require the necessity of knowledge with a specific detection threshold size. Therefore, if the detection threshold is drawn from between a small range of 500 to 1000 ms each time it is checked, the meaning will make it difficult to implement a program that jumps between values and does not "travel" in time. What's more, this program could theoretically be useful only in arcade games, in which the player can take a break between successive short sequences. In this case,

Table 1 The number of variables found after the Cheat Engine performed subsequent scans for a statically and dynamically changed address

No change of variable address				Dynamic change of variable address			
Scan 1	Scan 2	Scan 3	Scan 4	Scan 1	Scan 2	Scan 3	Scan 4
91	1	1	14231	3	0	0	12144
2	1	1	13665	6	0	0	13713
12	1	1	13726	2	0	0	13923

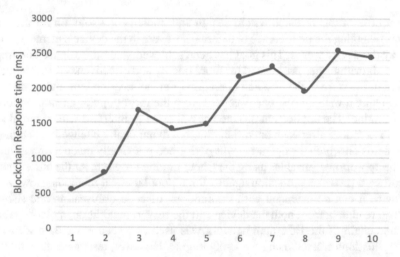

Fig. 5 Blockchain response times for subsequent attempts to download and change data

the player could perform the arcade sequences when the time is slowed down and wait a while when the time is accelerating, but for values below 1 s, it will not be in practice, useful (Fig. 5).

The blockchain has been tested for contract processing speed and for detecting interference in its memory. The tests were carried out in a local environment, and therefore the real results that would be carried out on a normally operating blockchain could be significantly different, most likely the time received would be greater. In addition, blockchain was launched on Windows at the Windows Subsystem for Linux layer. The response time was measured in the game and it is the time from the start of the value retrieval transaction to receipt of confirmation of its change from the block chain. The times obtained are shown in Table 2. They ranged from about 0.5

Table 2 Blockchain response times for subsequent attempts to download and change data

	1	2	3	4	5	6	7	8	9	10
Response time [ms]	543	2145	1675	1402	1476	786	2297	1945	2523	2428

to even 2.5 s. A performance test similar to those described in previous paragraph was carried out using Cheat Engine detected all attempts to interfere with memory and prevented them by overwriting the changed value by the one taken from the blockchain.

5 Conclusion and Future Work

The purpose of this research was to design and implement a solution for the Unity engine, detecting the interference of external software in the device's memory, and in the way the device works. This goal has been achieved, and the designed and implemented algorithms has been tested for functionality and effectiveness. To conduct an attempt to interfere in the data and operation of an application built using the Unity engine using a written tool, one was used from the most popular cheat programs— Cheat Engine. The application detected any attempt to interfere with external software. In addition, after each operation, the tool dynamically changed the addresses of the fields on which the operation was performed practically made it impossible to find the appropriate variables using the Cheat Engine program. At the current stage, the software provides protection against three popular cheat methods. However, in the future it can be extended with the function of code obfuscation and securely save and read data from both the device and the server. In addition, a blockchain set as a side chain of the Ethernum network was used as a secure database for games and applications created using the Unity engine. However, tests have shown that at present due to its limitations in the form of the number of transactions processed in each second and the amount of time needed to execute transactions, using it only for the purposes of data security of the game does not make sense. The exception may be a situation in which micropayments in the game are implemented using cryptocurrencies, because the number of operations is then many times smaller and on each of them the application developers earn, which reimburses the costs they have to bear for the maintenance of the blockchain.

References

1. Feng WC, Kaiser E, Schluessler T (2008). Stealth measurements for cheat detection in on-line games. In: Proceedings of the 7th ACM SIGCOMM workshop on network and system support for games. ACM, pp 15–20
2. Heo GI, Heo CI, Kim HK (2015) A study on mobile game security threats by analyzing malicious behavior of auto program of clash of clans. J Korea Inst Inf Secur Cryptol 25(6):1361–1376
3. Bremer J (2013) Automated analysis and deobfuscation of android apps & malware. Freelance Security Researcher
4. Consalvo M (2007) Cheating: gaining advantage in videogames. Massachusetts Institute of Technology, pp 5–8

5. McGraw G, Hoglund G (2007) Online games and security. In: IEEE Security & Privacy, vol 5, no 5, pp 76–79
6. Joshi R Cheating and virtual crimes in massively multiplayer online games. Royal University of London
7. Tolbaru SA (2011) Cheating in online video games, University of Copenhagen, Datalogisk Institut, 15 Aug 2011
8. Yeung SF, Lui JCS, Liu J, Yan J (2006) Detecting cheaters for multiplayer games: theory, design and implementation 1178–1182
9. Yan J, Randell B (2005) A systematic classification of cheating in online games. In: Proceedings of 4th ACM SIGCOMM workshop on Network and system support for games. ACM
10. Park JK, Han ML, Kim HK (2015) A study of cheater detection in FPS game by using user log analysis. J Korea Game Soc 15(3):177–188
11. Ahn D, Yoo B (2017). A study of cheating identification and measurement of the effect in online game
12. Raymond E (1999) The case of the quake cheats, Unpublished manuscript
13. Baughman N, Levine B (2001) Cheat-proof playout for centralized and distributed online games. In: Proceedings of the twentieth IEEE INFOCOM conference
14. Li K, Ding S, McCreary D (2004) Analysis of state exposure control to prevent cheating in online games. In: The 14th ACM international workshop on network and operating systems support for digital audio and video (NOSSDAV)
15. Cano N (2016) Game hacking: developing autonomous bots for online games. No Starch Press
16. Olleros FX, Zhegu M (2016) Research handbook on digital transformations, Edward Elgar Publishing
17. Swan M (2015) Blockchain: Blueprint for a new economy, O'Reilly Media
18. Zyskind G, Nathan O (2015) Decentralizing privacy: using blockchain to protect personal data. In: Security and privacy workshops (SPW), 2015. IEEE, pp 180–184
19. McDonald J (Valve), Using deep learning to combat cheating in CSGO

Approaches to Business Analysis in Scrum at StepStone—Case Study

Piotr Forowicz

Abstract Wide adoption of Agile has significantly impacted the sphere of *business analysis*. Although need for *business analysis* is widely accepted, it is often unclear how it should be applied. Opinions about the role of a BA (Business Analyst) in Agile teams vary greatly—from negation of its existence to acceptance of a BA as a team member, working side by side with programmers. At StepStone, since our switch to Scrum in 2014, we have experimented with several models of team organization, with or without dedicated BAs. In this paper, the two most prominent cases will be presented: when the BA was a full-time member of the development team, and when the BA's role was distributed between other team members—developers and the Product Owner. Advantages, disadvantages and the transition process will be discussed. In addition, techniques will be proposed which may help in development of necessary analytical skills and transforming team organization - based on our experience.

1 Introduction

We live in a rapidly changing world. Companies must survive in highly competitive business environments. Time to market—length of the whole software change life-cycle is expected to be reduced by IT. Agile methodologies that promise frequent (even almost immediate) delivery and short lead time, have become the standard. At the same time, we see that Business Analysis and Requirements Engineering are key areas for success, understood as delivering useful, valuable products in reasonable time and cost ("subjects with traceability performed on average 24% faster on a given task and created on average 50%", while traceability means in this case maintaining a link from requirement to solution [1]).

Requirements Engineering is a very important part of Software Engineering. When not properly done, it can endanger the success of the whole endeavor. Many

P. Forowicz (✉)
StepStone Services Sp. z o.o., Warsaw, Poland
e-mail: piotr.forowicz@stepstone.com

© Springer Nature Switzerland AG 2020
A. Poniszewska-Marańda et al. (eds.), *Data-Centric Business and Applications*,
Lecture Notes on Data Engineering and Communications Technologies 40,
https://doi.org/10.1007/978-3-030-34706-2_11

211

companies and teams have to solve the issue of Requirements Engineering in Agile-oriented ecosystems.

Let's take a look how we at StepStone have dealt with the issue.

2 What Is Business Analysis

BABOK [2] defines business analysis as:

> practice of enabling change in an enterprise by defining needs and recommending solutions that deliver value to stakeholders. Business analysis enables an enterprise to articulate needs and the rationale for change, and to design and describe solutions that can deliver value.

My personal short definition of this profession or role is "The one who knows *What*". To elaborate—"The one knowing *What* should be done and why". While managers focus on "*When*" and programmers on "*How*", BAs focus on "*What* ". In this case, "*What*" should not be understood only as a set of requirements or features (of e.g. software), but rather as a deep understanding of the business need and change required to satisfy that need.

2.1 Agile Manifesto Versus Business Analysis

The process of requirements elicitation and analysis is commonly called "*refinement*" in the Agile world. Searching for explicit roles in the Agile Manifesto (Manifesto for Agile Software Development [3]) is in vain. There are, however, some clues often misunderstood as encouraging the omission of the analysis stage or even the requirement and need analysis at all. The Agile Manifesto says[1]:

> Working software over comprehensive documentation

Does this mean that you should not document? No. But it is often quoted as justification for a lack of any documentation whatsoever. Managing requirements, needs and assessing alternative solutions is not against Agile. I would even go so far as to say that it is at the core of Agile, as focus on delivering "*working software*" means that we have to understand what "*working*" really means in a particular case.

Let's quote two of the twelve Principles behind the Agile Manifesto [4]:

> Continuous attention to technical excellence and good design enhances agility

> Simplicity–the art of maximizing the amount of work not done–is essential

Both emphasize the part we call Business Analysis—good design and focus on real need require precise understanding of that need.

---

[1] Quotes from the Agile Manifesto should always be understood according to the general principle: while there is value in the items on the right, we value the items on the left more.

2.2 Scrum Versus Business Analysis

The Scrum process relies on the Scrum team: The Product Owner who is the (one and only) gate to the team, Scrum Master and Scrum Developers (Development Team). Other supporting roles are outside the team, including Customers, Domain Experts and other Stakeholders.

The Scrum guide recommends building Development Teams comprising of 3–9 members—this number does not include the SM (Scrum Master) nor PO (Product Owner). Scrum values self-organizing and independent teams. This does not mean that the role of the BA *is* or *is not* within a team. Team members are not uniform plastic people, and can specialize in database design, frontend or business analysis. Some common myths are:

- The Development Team, as its name would suggest, comprises developers = programmers only
- All team members should be interchangeable, thus eliminating the BA with his unique role
- The PO is responsible for saying "*what*" and "*when*" to do, so what is the BA for?

Is this really written in the Scrum Guide? Let's take a look. The Scrum Guide [5] says that a team should be:

> self-organizing and cross-functional

and

> Cross-functional teams have all competencies needed to accomplish the work without depending on others not part of the team.

In "The Development Team" chapter we can also find:

> Scrum recognizes **no titles for Development Team members**, regardless of the work being performed by the person;

> Scrum recognizes no sub-teams in the Development Team, regardless of domains that need to be addressed like testing, architecture, operations, or business analysis;

and

> **Individual Development Team members may have specialized skills and areas of focus**, but accountability belongs to the Development Team as a whole.

This may seem contradictory. So, team members can specialize and focus on specific areas, but we shouldn't name their roles…

3 Two Practical Approaches

3.1 Approach #1: The BA as a Member of the Development Team

In the first discussed approach, the development team has specialized members, all of them focused full-time on their specific role. This includes:

- The Product Owner,
- Scrum Master,
- Quality Assurance Engineer(s),
- Developers with Team Leader
- Business Analyst(s).

Obviously, this implies teams being rather numerous—in our case it was about 12 people per team.

3.1.1 Business Analyst Duties

In approach #1, the BA closely operates with the PO, with daily alignment meetings (typically remote) and face-to-face workshops c.a. once a week. The BA uses the remainder of his time for refinement (aka backlog grooming), acceptance criteria writing, also documenting work done by the team and acceptance testing on behalf of the PO. The BA is fully responsible for refinement meetings, during which the development team estimates stories. When needed, additional technical refinement meetings are held with developers. Developers communicate with the PO a lot less than with the BA. The BA is located on site with the development team, while the PO is located on site with business.

3.1.2 Impact on the PO Role and Team Size

Having the full-time support of a business analyst, sometimes even two, Product Owners focus more, as one of them put it, on "Foreign Affairs". This means communication with decision makers, lobbying, but also solving problems of end users, visiting them on site and convincing them that we care, as well as market analysis, strategic planning and roadmapping.

3.2 Approach #2: BA Role Distributed

In the second of the presented approaches, the development team is more uniform. We have:

- The Product Owner,
- Quality Assurance Engineer(s),
- Developers with Team Leader (50%).

Yes, this means most team leaders have two teams. Former Scrum Masters and BAs play now a TL (Team Leader) role or support several teams.

3.2.1 Distribution of Business Analysis

In approach #2, the team relies more on the PO and tickets have to be more detailed from the beginning. As there is no dedicated business analyst, refinement process duties must be distributed within a team. Most business system analysis duties are assigned to senior and regular developers. Some teams assign these tasks during daily ("stand up") meetings, while others dedicate a separate meeting or part of the refinement meeting. The latter seems to be most effective, as people are more consciously undertaking tasks, and so are more dedicated to them. The PO is still the source of knowledge, but developers in an analyst role (we call it "BA responsible") also use other information sources in business and IT.

More analytical tasks for the PO imply a need for the reduction of the area he can cover. We experimented with Product Managers ("smaller POs"), finally splitting teams by half.

In this approach, teams are additionally supported by very few BAs located in the architecture team. These BAs are now more advanced generalists than system analysts. They are involved in large, cross-team projects and serve as coaches and trainers.

3.3 Refinement Process

The most effective process worked out is common for both approaches:

1. Somebody, most often the PO, prepares a request (ticket, story)
2. The PO accepts it to backlog or closes it
3. If there is no dedicated BA in the team, the person responsible for analysis (BA responsible) has to be assigned
4. The BA responsible, together with the PO, business and other stakeholders if needed, refines the story
5. Optional: technical approach(es) may be prepared
6. The Team reviews the story, verifies DOR (Definition of Ready) compliance and estimates effort on meeting
7. The Ticket is ready for development.

3.4 Tools

We store team backlogs in Atlassian Jira. Requirements can be described by a canonical User Story with Acceptance Criteria schema, but other approaches are also utilized. Some teams experimented with BDD (Behavior-driven development) and TDD (Test-driven development), which were not commonly adopted. Definition of Done and Definition of Ready are not strictly enforced. When assessing if the work was done or not, we rely more on informal communication and agreement between the PO and the development team.

4 Reasoning for Change

Several development teams moved from approach #1 to approach #2 after ca. 1, 5 year, about 3 years ago. The main reason for the change was wanting to have whole teams dedicated to the business goal, and to achieve higher transparency. In other words, to emphasize and increase the developer's functionality ownership sense. The change was well-perceived by business, however, approach #1 is also used successfully by other teams. We can't just say that approach #2 is the silver bullet; rather the advantages and disadvantages presented below should be considered to assess the right approach for a specific situation.

5 Comparison

The main factors differentiating the presented approaches are enumerated in Table 1. The comparison is for a near ideal case, when both approaches are implemented properly. The most important and less obvious factors are discussed in more detail below.

Programmers ownership sense, attitude: In Approach #1, a decrease of programmers' functionality ownership sense was observed. When the BA and PO prepared detailed acceptance criteria, programmers weren't involved enough in the process, thus losing a sense of ownership both for requirements and implemented features. What should be emphasized is that this lack of involvement was observed also when programmers were encouraged to take part in analysis. Even requirements very well written by PO+BA were often challenged as not precise enough. In Approach #2, programmers identify themselves much more with refined user stories (requirements).

Transparency as perceived by business: Not surprisingly, Business people like to have direct impact on the programmer's work. Business analyst (or any other role in between) creates longer communication lines; also, often the BA asks many in-depth questions that are not perceived very well in organizations with a very high

Table 1 Advantages and disadvantages of presented approaches

	Approach #1 BA role explicit	Approach #2 BA role distributed
Size of team	c.a. 12	c.a. 6
Programmers ownership sense	Lower	Higher
Transparency as perceived by business	Lower	Higher
Ease of implementation and balancing of roles	High, due to specialized roles	Lower, training and coaching needed for most teams
Team focus on requirements & needs refinement (analysis)	High	Lower, requires constant care
Final quality of requirements & docs	Higher	Lower
Distribution of business knowledge in team	Concentrated: PO, BA—less so the rest of the team	More even, although still senior staff are more knowledgeable than juniors
Flexibility: task assignment, switching members between teams	Lower	Higher
Risk related to absence	Higher—a missing person can significantly hinder work	Lower
Training requirements	Specialized per role	Programmers have to develop themselves in areas they do not always like
Business travels needed	Mostly BA or PO	Most programmers (at least senior and mid-level)

"can do" attitude. Often perception of developers who do unneeded work, but later with the utmost devotion stay after hours to correct their epic "Fast fail", is better than of those who focus on analysis and then effectively program only really needed functions which work well from the outset. This may be because it is very difficult to measure development progress other than in spent hours. Last but not least, everyone nowadays understands the need for programmers, while the need for BAs is not as obvious.

Ease of implementation, balancing of roles: Approach #1 is much easier to implement, as people play specialized roles in accordance with their experience and job position. Most uniform teams, as in Approach #2 where team members need to be universal, have difficulty with balancing time devoted to various development phases. Combining the roles of programmer and business analyst in one physical person is also difficult due to the different personalities required. Not surprisingly, developers are more interested in development than in analysis. This impacts not only day-to-day working behavior, but also personal development (**Training Requirements**). Programmers are keen on new libraries, tools, languages, but are not so much interested in business analysis techniques, requirements engineering and other factors impacting the understanding of business needs, such as benchmarks, compliance, competition on the market etc. The "fun factor" has to be taken into account if we want to attract the best talents!

The same imbalance may be observed for other roles, especially those focusing on long time outcome, such as the Scrum Master. Although teams wish to improve, without a Scrum Master in their vicinity they focus more on daily firefighting than investing in better future process. An important factor impacting the imbalance is pressure on development tasks from outside the team. While everybody understands that programming tasks have to be undertaken, analysis and quality assurance are not as commonly appreciated. This pressure is much easier to control within a team when the BA is a specific role—he simply cannot code so his focus on future is much harder to be cannibalized by "business criticals" (urgent bug fixes).

The most important criteria: **quality and amount of delivered work** were unfortunately not measured during the change. However, subjective perception of change from Approach #1 to #2 was positive, so delivery most likely was at least at a comparable, if not better, level.

6   Transition—BA Skills Improvement Program

When teams were divided and BA responsibility put on developers' shoulders, the need for assessment and improvement of required BA skills emerged. The improvement program started from assessments:

- analysis of PO's and TL's challenges
- assessment of programmers' skills.

Analysis of the above leads to long term activity comprising training, coaching and hand-to-hand work. The program was completed by two former BAs and took about two months per a team.

6.1   Analysis of PO Challenges

Half a year after moving to Approach #2, POs and TLs (22 persons in total) were asked open questions about issues affecting the performance and productivity of their teams. There were 26 issues reported related to business analysis. Due to open questioning, it is not possible to quote all the answers, the most popular being:

- a lack of a big picture (8 times),
- a lack of a BA in the team (5 times),
- not enough analysis,
- overly technical refinement meetings.

6.2 Assessment of Programmers' Skills

After teams split, we assessed the necessary analytical skills of team members using a custom competency model, a simplified version of IIBA one [6]. Analytical skills of every team member were self-assessed with the help of a Team Leader in the 1–5 range, with 1 being a novice, and 5 proficient (as a benchmark: the expected grade for an entry level BA would be about 3).

The average result across all development teams was close to 2.3. There were huge differences between teams observed, from ca. 1.45–3.0. Such values clearly indicate that only a few teams were then able to conduct business analysis related activities effectively enough, based on the skills of their members. This led us to the conclusion that improvement activities are needed.

6.3 Improvement Activities

In order to solve the aforementioned issues and improve analytical skills, improvement activities were also proposed and discussed during interviews. The following activities were chosen based on feedback: training, joint workshop with business, refinement assessment. Other proposals gained lower acceptance, some of them were introduced later but with lower priority. Examples: adjust and enforce DOR (Definition of Ready) and DOD (Definition of Done), BA best practices promotion, rotational roles (e.g. BA or SM).

6.4 Training

Training focused on showing programmers the value of *work not done*, and focusing on *doing right things*, encouraging team members to openly communicate, understand their goals and the PO's needs, also to challenge the PO's overly technical requirements. Two half day workshops proved to be enough to impact mindset and create space for other activities. At the teams' request additional trainings were provided, focusing on specific aspects such as: "What is the user story" or "How to create good acceptance criteria".

6.4.1 Joint Workshop with Business

In an ideal case, the joint workshop comprises:

- a lecture led by businesspeople
- observation (shadowing): assisting business staff in their typical daily tasks.

Observation has to be carried out in small groups (one businessperson + a maximum of two technical persons observing), thus including all team members was not always possible. This technique gives a huge boost to understanding and communicating with business. It not only provides knowledge, but also creates very important bonds.

6.4.2 Refinement Assessment

In this activity, an experienced BA takes part in refinements of the team in order to observe and propose improvements—both during retrospective meetings and in daily work.

6.4.3 Other Activities

The BA supporting the team during the improvement program spent about 20% of his time with the team. In addition to the activities already mentioned, he often helped solve issues and was carrying out pair analysis with team members in order to create user stories to be used later as a reference. He also suggested process improvements and tooling.

6.5 Results

The program began on 2017.07.04 with one team, and by the end of 2017 most teams had at least started the program. As such, soft skill trainings do not increase effectiveness immediately, so we were not able to measure direct impact. However, Table 2 shows that organization improves over time.

Table 2 Sprint goals achievement statistics

Year	2017		2018												2019		
Month	11	12	1	2	3	4	5	6	7	8	9	10	11	12	1	2	3
Sprint cancelled		1		1						1		2					
Sprint goal missed	2	1	12	9	8	10	12	11	16	14	13	11	11	9	6	8	2
Sprint goal achieved	3	2	10	15	7	15	25	27	32	24	26	34	23	27	30	20	5
Finished sprints	5	4	22	25	15	25	37	38	48	39	39	47	34	36	36	28	7
Success ratio (%)	60	50	45	60	47	60	68	71	67	62	67	72	68	75	83	71	71
Success ratio (graph)																	

The results presented here come from the internal ticket management system (Atlassian Jira). Every sprint is managed here as a ticket. Finished sprints achieve a status: Cancelled, Goal Achieved, Goal Missed. All development teams are included in this statistic. Excluded are teams which do not work in sprints, e.g. architecture, and database engineers who rather support other teams. The number of "pure" development teams changed over time from 12 to 20. The number of analyzed backlogs (22) is higher than the number of teams. This is caused by the fact that at StepStone, there is no 1:1 relationship between backlog and a team. For example, there were specific "task forces", formed with the same team members as regular teams, having their own backlog. Also, there are legacy backlogs managed separately, as they are related to a specific area, so one regular team may be responsible for more than one backlog.

The main reasons for a missed sprint goal were underestimated effort and an unclear goal—both cases are clear examples of improper requirements management. Thus, improvement of the sprint goal achievement ratio can be interpreted as proof of an increase in business analysis skills.

7 Other Considerations

Programmers often expect very technical input from the PO. Many POs gladly fall into this trap, while not having enough technical understanding. It is much better if the PO presents the goal and business need (e.g. in the form of a well described epic) letting the team choose the solution and prepare detailed requirements (e.g. user stories with acceptance criteria). We were able to reduce effort by 2–3 times in real cases, both in the analysis and development phase, following this schema.

Teams are strongly integrated internally. This is good from a single team effectiveness point of view, but we still struggle with team coordination. Scrum is not scalable. It works well inside the team, but unfortunately does not help with integrating teams. Don't expect smooth introduction of scrum in large organizations—synchronization of work between teams is a real issue that has to be addressed. BAs can play an important role in aligning teams—in both presented approaches.

When performing analysis of difficulties and challenges POs face in their daily work, it was observed that most POs previously working with BAs feel a lack of their support after switch to Approach #2, while others without such experience do not. Worth noting is that the latter—not explicitly interested in having a BA in the team—often complain about issues in the BA area of responsibility, such as a lack of a big picture, unresolved dependencies with other teams and backlog quality.

8 Summary

In the paper, two practical approaches to the business analysis process and team organization were presented and compared. Both have advantages and disadvantages, which were discussed. Several activities which may help to build business analysis skills and awareness were proposed. Reviewing them may help organizations to choose the most effective approach.

Our analysis showed also that there are a tremendous number of opinions and articles available, though it is very difficult to find any numbers and scientific papers discussing the Agile development process. Such research can be carried out only by universities, as in business we are never able to create the same software many times, using competing approaches, in a controlled environment. Also, measuring over a long period of time in business organizations is hard, as structure changes very often in order to achieve the required flexibility.

9 About the Company

StepStone helps people to find their dream job and companies to find the right employees. As one of the leading job platforms in the Western hemisphere, StepStone connects job seekers with suitable jobs and employers through the use of intelligent technologies.

The entire StepStone Group with all its jobsites employs over 3,300 people and is represented in more than 20 countries worldwide. Founded in 1996, the company is a subsidiary of Axel Springer SE, one of Europe's leading media companies.

The StepStone websites offer direct access to well-qualified professionals from the most in-demand professional groups such as IT specialists, engineers and technicians, as well as professionals in finance, sales and marketing. It is designed to find the "perfect match" in all important occupational fields at every career stage.

StepStone reaches millions of candidates—putting it streets ahead of other job platforms. Thanks to high rankings on Google and other search engines, effective banner advertising, major collaboration partners and a strong presence on various social media networks, it is almost impossible for online jobseekers to overlook StepStone. The StepStone JobApp for iPhone and Android also ensures that the growing number of jobseekers using mobile devices can access our services as well.

The development organization plays an important role within the whole StepStone Group. The development teams support and extend software used by various job portals and other classified ads sites throughout the EU and elsewhere. Since the requirements and needs of the different job markets vary from country to country and from location to location, developer teams must take into account the different business processes and needs in their daily work.

10 About the Author

An experienced Business Analyst, working for StepStone Services since 2014, first as an Agile Business Analyst in one of the development teams, responsible amongst other duties for new responsive listing design. Now a Business Architect in a Step-Stone Architecture Team, supporting development teams in analysis skills and process improvement, analysis for larger projects, creating architecture standards.

References

1. Mäder P, Egyed A (2015) Empir Softw Eng 20, 413. https://doi.org/10.1007/s10664-014-9314-z
2. A guide to the business analysis body of knowledge v3 (BABOK Guide). International institute of business analysis, Toronto, Ontario, Canada. ISBN-13: 978-1-927584-03-3, available also online on: https://www.iiba.org/
3. Agile manifesto. http://agilemanifesto.org/
4. Principles behind the Agile Manifesto. http://agilemanifesto.org/principles.html
5. Official scrum guide. https://www.scrumguides.org/
6. Business analysis competency model v4 (2017) International institute of business analysis, Toronto, Ontario, Canada. ISBN-13: 978-1-927584-06-4

Development Method of Innovative Projects in Higher Education Based on Traditional Software Building Process

Rafał Włodarski and Aneta Poniszewska-Marańda

Abstract Numerous attempts were taken to define criteria against which to evaluate and measure software project success. The complexity that surrounds it has been recognized by companies, however the scholarly world is yet to follow. In this article, three dimensions of success have been elicited basing on prior industrial studies: project quality, project efficiency as well as social factors (teamwork quality and learning outcomes). Investigation of their assessment in a commercial context enabled the authors to determine and adapt a set of metrics and measures that constitute a structured evaluation scheme for projects developed by student groups. The proposed assessment approach can be applied by researchers appraising applicability of software development processes in an academic setting or carrying out comparative studies of different methodologies. Furthermore, professionalizing appraisal of student project outcomes can contribute to closing a gap between workforce's expectations towards new graduates and the effects of their university education.

Keywords Software projects evaluation · Project success · Project quality · Project efficiency · Capstone project · Academic context · Higher education · Curricula

1 Introduction

Over the course of history, the approach to evaluation of students' work has greatly evolved. It begun as straightforward grading systems [1, 2] that assessed level of assimilation of knowledge and shifted to project-based pedagogy that favours active exploration of real-world challenges and privileges soft skills development. While

R. Włodarski · A. Poniszewska-Marańda (✉)
Institute of Information Technology, Lodz University of Technology, Lodz, Poland
e-mail: aneta.poniszewska-maranda@p.lodz.pl

R. Włodarski
e-mail: rafal.wlodarski@edu.p.lodz.pl

© Springer Nature Switzerland AG 2020
A. Poniszewska-Marańda et al. (eds.), *Data-Centric Business and Applications*,
Lecture Notes on Data Engineering and Communications Technologies 40,
https://doi.org/10.1007/978-3-030-34706-2_12

project success is now regarded as a multidimensional construct, no framework that evaluates its different facets in case of students' work has been published.

Although studies devoted to the definition and measurement of success of industrial software development proliferate [3–5], in an academic context the same subject attracts little attention and focused on the source code by means of mining the repository data [6, 7]. The motivation in undertaking this research was to challenge how student projects are evaluated, draw on assessment criteria used for commercial IT deliverables and translate it into an academic grading context that would resonate with the demand and trends set by the ever-changing software industry. Based on these goals, the authors state the following research questions:

Research Question 1: What are the success factors in software development? or What constitutes a successful software project implementation and how can the success factors map to an academic setting?

Research Question 2: What metrics and measures are used in industrial software development to evaluate the success of a systems project and the process followed?

Research Question 3: Which metrics are pertinent to an academic setting and how to adapt them to the particularity of student projects?

In order to answer the above questions a multi-staged literature review helps to identify three areas of focus for evaluation of projects developed by students and a set of metrics, and measures for their assessment:

- *Project quality* (source code and product quality).
- *Project efficiency* (resources utilization and team productivity).
- *Social factors* (teamwork quality, learning outcomes).

In this paper we refine the earlier established metrics, categorize them along the proposed success dimensions, provide necessary adaptations for an academic setting and generalize them so that they can be applied to a broad spectrum of student software undertakings.

The implication of this paper for the Software Engineering community is two-fold and our results are presented in such a way that the success assessment scheme can be used in its parts or integrity:

- as a reference to monitor and evaluate the success of students' work along three dimensions,
- to evaluate or compare a software process in an academic setting.

The paper is structured as follows: the literature review approach is discussed in section two, while the three dimensions and their measurement methods are detailed in sections three, four and five respectively. Threats to validity and a discussion part conclude the article in section six.

2 Definition of Success Dimensions

The two-step literature review that underpins this study seeks to identify the different facets of software project success and establish methods of their evaluation. The search included electronic databases and conference proceedings using the following sources:

- ACM Digital Library,
- IEEE Xplore,
- ISI Web of Science,
- ScienceDirect—Elsevier,
- SpringerLink,
- Wiley Inter Science Journal Finder.

The search strategy behind each phase of the investigation was based on a combination of keywords that explore a specific subject (e.g. success factors, quality improvement) in a given context (Software Development, Extreme Programming) for a particular type of study (literature review, industrial case study); different associations between keywords used are presented in Fig. 1.

The first step of the investigation sought to define what constitutes a successful project implementation in a commercial context and served as a basis for defining the areas of evaluation for student undertakings based on 200 primary papers. One of the most widely known concepts in this area is the "iron triangle" which translates into respect of the project's time, cost, and quality constraints [8–11]. However, this long-established model has been denounced for its sole focus on the project management process [11–13] and and a need to recognize a distinction between the success of a project's process and that of its product [12, 14] was expressed. Numerous further extensions of the success factors have been proposed, including effective project team functioning [12, 15] and benefits of different project stakeholders [10, 11]. McLeod et al. [16] summarize various criteria proposed in project management literature along three axes: Process Success, Product Success, and Organizational Success. The first two dimensions are applicable to any type and setting of a project whilst the third one necessitated a mapping to an academic context and its stakeholders (Sect. 5).

The second stage of the literature review targeted metrics and measures that are commonly used to evaluate different aspects of systems development success. Many relevant studies were selected and most of them were retained as part of the primary study selection based on the following inclusion criteria:

- availability of calculation method and data acquisition procedure,
- description of the reasons for and effects of using the metric,
- applicability of the metric at the team or company level,
- possibility to collect and use the metric in projects of any scope, size and complexity.

The objective was to collect as many eligible papers that would provide a broad vision on assessment in the software industry given that the third research question

	Population	Context	Intervention
RQ1	business value OR business impact OR project success OR project management success factors OR success factors OR process performance OR process improvement initiative OR SPI OR human factors OR KPI OR success dimensions	software development OR Agile OR Scrum OR XP OR Extreme programming OR Lean OR plan-driven OR waterfall model OR software process OR life cycle OR Software Engineering OR Industry Project* OR governance OR management	research OR analysis OR measure* OR evaluat* OR compar* case stud* OR Industrial case stud* OR stud* OR systematic review OR literature review OR systematic literature review
RQ2	metric* OR measurement program OR team management OR high-performing teams OR teamwork effectiveness OR effectiveness OR efficiency OR performance measures OR quality Improvement OR quality OR goal-question-metric OR transition OR transformation OR improving	software development OR development methodology OR Agile OR Scrum OR XP OR Extreme programming OR Lean OR traditional software development OR plan-driven OR waterfall model OR software process OR life cycle OR software engineering OR industry project*	measur* OR assess* OR evaluat* OR compar* OR analysis OR case stud* OR Industrial case stud* OR stud* OR systematic review OR literature review OR systematic literature review OR qualitative OR quantitative OR empirical

Fig. 1 Keywords combinations (values from different columns were joint by an OR statement) underpinning the literature review for Research Question 1 and Research Question 2

was highly restrictive. Many of the metrics used in the industry are driven by business or organizational objectives that are absent in an academic context. Moreover, they are often process-specific. The evaluation scheme proposed in this article consists of metrics and measures that are thought to be generic with regard to:

- *application granularity*: metrics are pertinent to different types of student projects (individual/group work) and can be collected over different periods of time (fortnightly, for the entire semester, sprint etc.),

- *suitability to different settings*: measures are not bound to a certain sub-domain of computer science nor a development process followed (plan-driven, agile, or absence thereof).

3 Project Quality

A research of Paul Ralph and Paul Kelly [17] that examines the dimensions of success in software engineering, yields 11 themes with project being the most relevant and central concept. A study executed at Microsoft [18] confirms the finding that data on quality is the most important to the stakeholders of a software development project. The proposed evaluation approach makes a distinction between internal and external quality and proposes dedicated measures for both.

3.1 Internal Quality

Researchers in computer science and commercial institutions have been prolific in defining software quality metrics. Its general notion encompasses an array of aspects the taxonomy of which was defined in the ISO/IEC 25010 international standard on software product quality [19]; it embraces facets pertaining to both functionality and technical traits. The focus of the upcoming section is on the latter, which is grounded in the source code base. A version control system, employed to a large extent in computer science related courses, can be used as an insightful source for calculation of structural code-level metrics and tracking data. Source code and the compliance with continuous integration practice are regarded as two major factors influencing internal project quality in the proposed evaluation scheme.

Complex code structures are proven to be difficult to understand and more likely to generate errors as compared to a well-designed module [20]. Complexity has a direct impact on the quality of a product, its maintainability and ease of troubleshooting. A common measure used in the industry is Cyclomatic Complexity as it proved to reveal insight on internal code quality [21].

3.1.1 Code Complexity and Size Measure

Cyclomatic complexity (CC) determines complexity of a program by counting the number of decisions (linearly independent paths) made in a given source code. It is a standard metric in the industry and as reported by McGabe Software Company [22], CC meets the three qualities of a good complexity measure, namely:

- it is descriptive, objectively measuring some quality—decision logic in the case of CC,
- predictive, correlating with some aspect—errors and maintenance effort,
- prescriptive, as it guides risk reduction—testing and improvement.

Further findings of the Software Assurance Technology Center (SATC) at NASA [23] concluded that it is a combination of the code size as well as the complexity that proves to most effectively evaluate quality. Source code of significant size and high complexity bears very low reliability. Likewise, software with low size and high complexity, as it tends to be written in a very terse fashion, renders the source code difficult to change and maintain [24].

These code quality metrics intend to evaluate similar characteristics which can be embodied by the term software maintainability. Industry reports show that practices regarding software maintenance and operations such as version control, modular and loosely coupled design, and code reuse are scarce among graduates [25, 26]. This further justifies the use of an assessment tool that targets best practices and skills development in that area. As putting in place and following multiple metrics for code evaluation can be cumbersome and deter their use by students, application of a single maintainability measure that consolidates different technical aspects of software is advised, minimizing the potential overhead.

3.1.2   Maintainability Ranking

SIG, a software management consulting company, in collaboration with TV Informationstechnik have developed a measurement model that maps a collection of source code metrics to all the maintainability sub-facets as defined by the ISO 25010 standard: *analyzability, modifiability, testability, modularity and reusability*. The technical quality model involves analysis of the following metrics [27]:

- *Lines of Code (LOC)*, as increasing volume of the codebase implies more information to be taken into account and entails more maintenance effort,
- *duplicated LOC*, as repeated, redundant code requires rework of all of its instances in case of deficiency,
- *Cyclomatic Complexity*, as the simpler the code is the easier it is to comprehend and test,
- *parameter counts*, as unit interfaces with many attributes can indicate bad encapsulation [28],
- dependency counts, as tight coupling impedes modifications to the underlying code.

They are collected at different levels of the building blocks of the codebase: units (e.g. Java methods), modules (e.g. Java classes) or components (e.g. Java packages) and subsequently aggregated via grand total or quality profiles [28]. This operation allows to correlate its outcome to ratings of properties for the entire software project: volume, duplication, unit complexity, unit size, unit interfacing, module coupling, component balance and component independence. Property ratings are mapped to the sub-characteristics of maintainability as defined by ISO 25010 by calculating weighted averages according to dependencies designated in Fig. 2: every cross in a given column signifies contribution of a system property to the given sub-characteristic (row).

	Volume	Duplication	Unit size	Unit complexity	Unit interfacing	Module coupling	Component balance	Component independence
Analyzability	X	X	X				X	
Modifiability		X		X		X		
Testability	X			X				X
Modularity						X	X	X
Reusability			X		X			

Fig. 2 In order to calculate each ISO 25010 maintainability sub-characteristic rating a weighted average of its contributing system qualities (marked by X) is performed; for example, reusability is affected by units' sizes and interfacing in the underlying code, hence its rating will be computed by averaging the ratings obtained for those properties

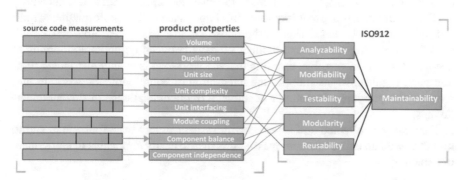

Fig. 3 Relationship between source code metrics, software properties and the ISO standard as defined by SIG maintainability model

This operation yields measurements which can be further combined to calculate ratings for the sub-characteristics and the general maintainability score for a given software system [29] on a scale from 1 to 5 (where higher number represents better performance). Figure 3 depicts the model in terms of relations between the metrics, product properties and maintainability characteristics defined by ISO standard.

For each property, a set of thresholds has been derived from an industry and open-source benchmark consisting of over 500 evaluations and encompassing 45 different computer languages to determine a quality ranking of the code in terms of a given property.

Multiple experiments have empirically validated the relationship between source code metrics for maintainability as defined in the SIG quality model and duration of maintenance tasks and indicators of issue handling performance [27]. A great advantage of this model of evaluation of internal quality aspect of students' undertakings is its ease of application and results' analysis. Its underpinning system properties were mapped to a set of 10 simple maintainability guidelines (as defined in Building Maintainable Software [30]) that need to be respected in order to produce quality code. Additionally, it comes along with a supporting tool—*TheBetterCodeHub*—

that checks compliancy of software projects against the guidelines via structural code-level metrics extraction embedded at the level of a GitHub repository.

3.1.3 Continuous Integration

First introduced by Booch [31], the concept of continuous integration aimed at avoiding pitfalls when merging code from different developers and in turn reduce the time and work efforts engendered by this process. CI rapidly became the industry standard and is now employed in the academia as well, as effective teamwork in student projects requires regular use of a version control system. A long-term study by Technical University of Munich (TUM) [32] reports that CI was perceived as beneficial by 63% of 122 students, whereas a mere 13% was not in line with that statement.

To measure adherence to this practice in a large-scale agile transformation in multiple companies, Olszewska et al. [33] propose a metric called "Pacemaker: Commit pulse" by counting the average number of days between commits and aiming at keeping it as low as possible, ensuring that the workload is evenly distributed. Furthermore, regular merges simplify the integration of team members' work, in turn reducing the pressure of meeting a deadline, a frequent challenge in student undertakings.

Data can be collected and assessed with respect to any time frame, e.g. sprint, month as well an entire semester making this metric easily applicable to any student undertaking.

3.2 External Quality

External product quality is quantified during the testing phase, by dynamic analysis of the product's behaviour as observed by its users [27]. Various kinds of software testing are essential to ensure software functional and technical product quality [28] however not all of them can be accommodated in a classroom. Whereas business projects comprise a dedicated testing phase, managed by a specialized team and running for several weeks, students usually perform only a rudimentary test campaign to make sure that the functionality is in place. That is followed by ad-hoc tests by the instructor at the end of the semester to evaluate his overall satisfaction level with the delivered product.

In order to address that gap, the authors of this paper propose to define metrics evaluating a subset of the product quality properties as defined by the ISO 25010 standard, prior to assignment implementation. That allows to incorporate it into the "Definition of done" criteria for Agile approaches or be used to guide a dedicated testing phase, when following a more traditional development lifecycle. In either case the metrics can serve as an evaluative measure of the project and ease the assessment of solutions for the instructor.

Moreover, students can be challenged to define the metrics themselves to infuse a real tester's thinking. Considering the array of possible projects implemented as part of Computer Science coursework and the information-rich taxonomy of product quality and quality in use, the measures can vary greatly depending on the type of the developed software system. For embedded systems projects or the ones based on CPU calculations, a particular focus can be put on performance efficiency in terms of time behaviour or resources utilization. Network Programming assignments could target the reliability feature with its availability and recoverability sub-characteristics.

In one of the authors' experiences, students whose objective was to implement an Artificial Conversational Entity supporting e-commerce catalogue browsing and product advisory, were asked to define a set of efficiency and usability metrics to assess the chatbots. They were required to:

- frame them according to the Goals-Signals-Metrics process,
- write test cases for evaluation,
- specify the method of metric procurement, the procedure of its collection and interpretation, and establish supporting tools used.

Once the teams documented their proposed metrics, a set of five most suitable measures to the context was chosen by the course instructors. Thereupon, metrics acquisition, calculation, storage and results representation method was specified and provided as a common reference for all the groups. That laid out a foundation for a "Jigsaw exercise" (Palacin-Silva et al. [34]) that was performed during one week of the testing phase when the students evaluated digital assistants developed by other teams according to a set of defined scenarios.

This approach enables assessment of the external quality aspect of software produced by students based on objective metrics and a large sample. The scores can be incorporated into the final grade for the projects, in turn facilitating solutions evaluation for the instructors.

4 Project Efficiency

From a classical project-management point of view the underpinning of a process's success is respect of underlying budget and time constraints. Indeed, a systematic literature review of 148 papers published between 1991 and 2008 [35] revealed that Effort and Productivity were defined as success indicators in 63% of studies of process improvement initiatives. Simply put, effort is reflected by the amount of time invested by the team and productivity by the output size in KLOC (kilo lines of code) [36] produced in the context of the development process. This paper proposes more comprehensive ways of evaluating a project in terms of efficiency and team productivity, as an incentive to produce large amount of code can have adverse effects on project quality (Fig. 4).

METRIC	CALCULATION METHOD	EVALUATED EFFICIENCY FACET
HUSTLE METRIC: FUNCTIONALITY/TIME SPENT	$HM = \sum_{i=1}^{n} Fp_i / \sum_{i=1}^{n} T_i$, where Fp_i: number of functional points of an artifact (task, module etc.) considered T_i: overall time spent by the team implementing the considered functionality	overall productivity of the team
PROCESSING INTERVAL: LEAD-TIME PER FEATURE	$PI = T_{ship} - T_{acc}$, where T_{ship}: timestamp when the feature is fully implemented and uploaded to a repository, T_{acc}: timestamp when the feature is accepted for implementation	efficiency of implementation process and capability to tackle encountered problems
WORK IN PROGRESS	$WIP = \sum_{i=1}^{n} Fp_i$, where Fp_i: function points of a task currently in progress	

Fig. 4 Project efficiency metrics, calculation method and aspect evaluated

4.1 Defining Measurements Units

The *Function Point Analysis (FPA)*, developed by Albrecht [37] introduces an antagonistic approach of measuring the functional size of a codebase; numerous standards and public specifications basing on the concept of *Functional Points* have been defined ever since. In parallel, other size-based estimation models emerged, such as *Use Case Points* [38] or story-based estimation in Agile techniques. While more comprehensive frameworks might yield precise results, they also require expertise and experience in the area, qualities that are scarcely among student groups. Function Point will thus be understood as an informed high-level estimation of an underlying piece of functionality (known as *Early Function Point Analysis*). Professors are encouraged to provide the estimates along with the specification of projects or assist and share their knowledge with students during the estimation process.

Adopting a time reference unit imposes that the students track their efforts spent on the assignment. Abrahamsson [39] proposes to collect time-related data for tasks under development with a granularity of 1 min and employing an excel sheet for that purpose.

Whereas this approach could be executed in a business context where a work schedule is clearly defined and more advanced supporting tools are at hand, the evaluation scheme proposes a precision of 15 min. Such a granularity is more sound given that students' work setting often involves efforts dispersed in time and little regard to time tracking.

4.2 Productivity and Efficiency Metrics

With the reference units for effort and time established, the ground is set for the definition of metrics used to evaluate the project efficiency dimension. The first one, suggested by Olszewska et al. [33] is "Hustle Metric: Functionality/Time spent", which measures how much functionality can be delivered with respect to a certain work effort. In order to calculate it, the amount of function points of a task, module or even an entire project is divided by the total quantity of time spent by the students during the implementation.

While the first metric gives an overall idea of how much output a project yields and the time needed for its completion, it is "Processing Interval: Lead-time per feature" [33] that gives insight into its flow. Processing interval of a feature is obtained as a subtraction of a timestamp when it is finished and uploaded to a repository (*Tship*) and a timestamp when it was accepted for implementation (*Tacc*). Monitoring turnaround time of features currently under development can reveal their technical or functional difficulties caused for the team; by analysing those of already terminated can shed light on the precision of an initial estimate.

In a study performed at Timberline Inc. [40], a related metric was used—*work in progress (WIP)*—which is an aggregation of function points of all features under development at a point in time. The authors observed that large amounts of work in progress induce more unidentified defects, that would be discovered eventually. In the context of student projects, WIP can be tracked in order to mitigate another baleful phenomenon: cherry-picking features that are most interesting or perceived as the simplest ones by the team leaving out the toughest ones to end of the term. Controlling WIP could reinforce a limited quantity of features to be implemented at a time, thus preventing the completion of low-priority and highly-appealing tasks before-all-else.

5 Social Factors

According to the study by Hoegl and Gemuenden [41] there are three leading factors that shape the success of innovative projects:

- team performance,
- teamwork quality,
- personal success.

While the efficiency aspect and its significance have already been covered, the following section focuses on teamwork as part of university classes and students' accomplishment. The objective of computer science higher education is to prepare corporate-ready graduates to be apt at programming and equipped with practical skills such as efficient collaboration and effective communication which can be developed while carrying out team projects. To assess growth in these areas, measurement methods of the teamwork quality as well as learning outcomes of students are proposed.

The evaluative tools described in this section can only be examined in a qualitative manner. The most straightforward and common way of doing so is through the use of opinion polls [32, 42, 43]. For the purpose of the proposed assessment scheme, it is suggested that all statements should be evaluated by students on a 4-degree Likert scale (as it leaves more space for nuance and omits a neutral answer): *strongly agree, agree, disagree, strongly disagree.*

5.1 Teamwork Quality

Teamwork quality is a measure of conditions of collaboration in teams; according to Hoegl and Gemuenden [41] it consists of six facets: communication, coordination, balance of member contributions, mutual support, effort and cohesion. Capturing any of these properties of cooperation within a group is a baffling task, hence the most representative of these characteristics is examined—cohesion.

5.1.1 Team Cohesion

Team cohesion is defined as the "shared bond that drives team members to stay together and to want to work together" [44]. As stated in [45] cohesion is highly correlated with project success, critical for team effectiveness [46], and leads to increased communication and knowledge sharing [47]. As cohesion emerges over time and its perception among the group variates as the project progresses [45], its investigation needs to be carried out periodically. There are no restriction on the time intervals but they should be frequent enough for the assessment to stay pertinent, e.g. every sprint or month.

In the context of Software Engineering, two dimensions of cohesion can be distinguished: attachment within the team (*social cohesion S* [45]) and attachment to the project (*task cohesion T* [45]). It can be further categorized according to granularity: at project member level (*Individual Attractions to the Group ATG* [45]) and team level (*Group Integration GI* [45]). These two distinction levels yield the following aspects of cohesion:

- GI-T: The team's attachment to the task,
- GI-S: The team's social connection,
- ATG-T: Individual attachment to the task,
- ATG-S: Individual connection to the team.

Carron et al. [48] developed *The Group Environment Questionnaire*, which measures all four aspects of cohesion via a survey. This questionnaire was devised for sports teams and proved to successfully assess team and individual perceptions of cohesion. While it requires some adjustments for computer science related projects, it is still applicable as demonstrated in a study at Shippensburg University [49]. Authors of the article used the following subset of adapted questions to evaluate team cohesion:

- Our team is united in trying to reach the goals of the course (GIT).
- Our team members have conflicting aspirations for the project's future (GIT).
- If members of our team have problems while working on the project, everyone wants to help them so we can get back on track together (GIT).
- Our team would like to spend time together once the course is over (GIS).
- Members of our team do not stick together outside of the course-related activities (GIS).

- I am unhappy with my team's level of desire to succeed (ATGT).
- This team does not give me enough opportunities to demonstrate my abilities and skills (ATGT).
- I do not like the collaboration style between the team members (ATGT).
- I do not enjoy being a part of the social activities of this team (ATGS).
- For me, this team is one of the most important social groups to which I belong (ATGS).

Wellington et al. [49], apart from employing G.E.Q. to compare the students' experiences with Plan-Driven and Agile Methodologies, also developed a survey permitting for further investigation of cohesion within a team. They focused notably on individual perceptions of the degree of "value" of each team member and closeness and mutual respect within the team. In the questionnaire, students evaluated every team member's impact on the overall project's success on a scale from 0 (*detrimental to project success*) to 5 (*critical to project success*). The second question required the students to hypothetically decide which team members they would include in their group while forming teams for a new project; a binary response reflected their intent to keep a team member (1) or not (0).

Although not mentioned in their study, thanks to the results of such a survey a professor can analyse the dynamics of a given team, find its weakest and strongest links, which would greatly contribute to a fair evaluation. Project quality is a team result, yet its contributors are graded individually in a university context; taking into consideration this metric could guide awarding higher scores to high performers in a team.

Further processing of the results of the first question—*inter-rater agreement*—yields a strong measure of the closeness of a team. Reciprocal relationships within the answers to the second question (pairs of team members that selected each other) give insight on the mutual respect within a team.

5.1.2   Team Morale

A common observation is that successful teams are happy and as such, they are efficient and produce quality results [34]. Measuring happiness has been gaining popularity in Agile software development frameworks as they emphasize teamwork and recognize its human aspect. The most common practice measures a happiness index by employing a Niko-niko calendar, also known as smiley calendar, where team members systematically rate their mood with a smiling, straight-faced, or frowning smiley. As appealing as that might be to students, in order to evaluate teamwork quality and personal success at the same time one should investigate team morale instead. Its classical meaning reflects a sense of common purpose and the amount of confidence felt by a person or group of people. In a commercial setting it is linked to job satisfaction, outlook, and feelings of well-being an employee has within a workplace setting which also resonate with conditions a university course should provide.

Van Boxmeer et al. [50] created a set of items to measure Team Morale in a military context and ever since, their work has been successfully used [51] to investigate causes and consequences of different morale levels within teams. They have also been adapted [52] to software engineering teams, whose members periodically indicate their agreement with the following phrases:

1. I am enthusiastic about the work that I do for my team.
2. I find the work that I do for my team meaningful.
3. I am proud of the work that I do for my team.
4. To me, the work that I do for my team is challenging.
5. In my team, I feel bursting with energy.
6. In my team, I feel fit and strong.
7. In my team, I quickly recover from setbacks.
8. In my team, I can keep going for a long time.

Each team member should rate the questions ideally halfway through the sprint (the most representative period of work performed) for agile methods and every three weeks for other settings. Team morale can be tracked all along the semester to probe the conditions of collaboration among the students. The average value for the team should increase over time whereas abrupt changes could indicate conflicts or periods of the project when the team struggled to meet course objectives.

5.2 Learning Outcomes and Skills

A final facet of project success is personal accomplishment of its participants. Although it might not be apparent to students, it is their learning outcomes and improved skills that are of paramount importance in that subject matter. Employers emphasize that both technical and soft skills are essentials for implementation of successful software projects. A study by Begel et al. [53] on struggles of new college graduates in their first development job at Microsoft finds that they have difficulties in teamwork and cognition areas. Brechner [25] suggests they should participate in dedicated courses in Design Analysis and Quality Code as part of their education in order to address the identified qualifications gap.

The proposed scheme evaluates student competencies divided into two categories, along with their their constituents:

1. *Software engineering skills*, that encompass: requirements elicitation, system design, data modelling, programming.
2. Non-technical skills, that encompass: communication, teamwork.

To increase the validity of the evaluation, the professor can contribute to the process by assessing the produced artefacts. Clark [43] has mapped learning outcomes and skills to corresponding assessment tasks as part of his study on student teams developing industry projects; its version enriched with evaluation tools described in this article is presented in Table 1.

Table 1 Learning outcomes and graduate skills—mapping of projects' outcomes, developed skills along with artefacts and proposed measures that can be used for their evaluation

Learning outcomes	Graduate skills	Assessment artefacts
Working in a team, students will be able to question a client/professor to extract and analyse the software requirements and present the analysis in a written report	Software engineering skills—requirements elicitation. Communication skills	Requirements documentation (UML diagrams, Functional requirements document, User interface specification). User stories
Having analysed the requirements, students will be able to prepare appropriate design documents while working in a team	Software engineering skills: - system design, - data modelling	System architecture diagram, Detailed design document, Database model, Data structure diagram, Entity-relationship model, Wireframes
Having prepared design documents, students will be able to construct and integrate a significant software system while working in a team	Software engineering skills: - programming, - version control. Team-working skills	Developed software evaluated along its internal and external quality aspects Repository (Pacemaker: Commit pulse metric)
Having developed a software system, students will be able to produce written technical and instructive documentation on the implemented solution	Communication skills	User manual. Installation guide
Students will be able to formulate a schedule for a team of people and individually and collectively manage their time	Team-working skills. Communication skills	Issue and project tracking software (Processing interval metric, Work in progress metric). Time reports. Risks document
Students will be able to work in a small team, planning effectively and be able to evaluate their own and peers performance at team and individual activities	Team-working skills. Communication skills	Issue and project tracking software. Team cohesion questionnaire

6 Conclusions and Discussion

This paper provides professors and researchers an approach for the evaluation of computer science project success that encompasses three dimensions, further divided into sub facets and addressed with a specific metric, measure or pedagogic tool. A description of the metrics and necessary adaptations for an academic context is provided, along with a discussion of the validity and appropriateness of the proposed measures for their intended use. The work described in this article has a number of implications.

With respect to practice, recognizing the multi-faceted nature of success mapped to specific metrics can help educators define a set of systematic evaluation criteria that can be used to appraise software project outcomes and consider the quality of team-work in group undertakings. Underlying such an approach is the notion that evalua-tion of students' work should replicate expectations that will be set towards graduates by their future employers. As many studies show [25, 26, 53], in order to efficiently perform in industry and respond to skills requirements of new software development professionals, more focus should be put on teaching operation/maintenance practices (version control, modularity, loose coupling, code reuse) as well as Human Factors such as communication and collaboration. Professionalizing the assessment process by incorporating the aforementioned aspects is one way of closing that gap.

The proposed evaluation approach contributes to the field of software engineer-ing in terms of empirical research methods. First off, it can serve as an analytical framework for scientists investigating the applicability of software developments methodologies to a university setting. It determines areas of focus for evaluation, which can underpin given hypotheses and suggests dedicated metrics and measures that can be used for their verification. Moreover, as it provides a set of measures and pedagogic tools that are course organization-agnostic it can be applied to com-pare different development methodologies in terms of the project outcomes and process-related attributes. Researchers interested in gaining a deeper understanding of different processes dynamics, including the observable characteristics between different development approaches, are encouraged to apply the proposed assessment scheme in the context of their studies. As continuous improvement of the systems development life cycle itself is a central tenet to software engineering community, we hope that the presented work contributes to this idea by extending the body of knowl-edge as well as enabling further empirical investigations by providing a re-usable evaluation approach.

While in this paper the focus is put on multi-dimensionality of project success, in practice not all criteria may be considered relevant or equally important on all student undertakings. The proposed evaluation approach indicates areas of assessment but leaves freedom to instructors on the choice of suggested measures. Furthermore, despite the fact that the metrics proposed in this article were selected because of their suitability to a university context, some limitations should be taken into account during their application and assessment. They pertain mostly to the metrics related to the efficiency and regularity of students' efforts as their work organization varies greatly from that of full-time developers. Dividing time between different courses and potentially an active social life poses a threat to validity of the following metrics: Processing interval: lead-time per feature, Work In Progress and Pacemaker: commit pulse. The authors' experiences show that commits to student repositories often cluster in time, come in focused bursts and are succeeded by a period of relative inactivity, which also negatively impacts updating of project tracking software.

This issue can be tackled and proved viable in authors' experiences when students follow a structured development process. Respecting artefacts and deadlines expected by a given methodology imposes regular work efforts and planning ahead. Instructors should share the schedule and due dates for deliverables in the beginning of the course and communicate on them continuously as the semester progresses. A tool linked to the traditional project development process which can further support regularity of students' efforts is formalized risks management. A given course can require students to capture possible interference with assignments and exams for other classes as part of a risk assessment document, which would then be subsequently updated and followed by the team's project manager or another designated member.

Thus far the assessment scheme was applied to two Master courses of a similar set-up (long term group projects over a complete semester) and research sample. Nevertheless, antagonistic systems development approaches were followed by students (Waterfall-like process and an adapted version of Scrum), thus contending its generic use in terms of software process. Further experimentation with the proposed work is planned at an additional university by utilizing a divergent study configuration in terms of project duration and student maturity with a goal to compare appropriateness of different software development methodologies to the course. Finally, the authors would like to verify if the professionalization of the assessment process favours increased learning outcomes and qualitatively explore which of the proposed metrics are the most beneficial for students in that regard.

References

1. Pierson G (1983) C. Undergraduate studies: Yale college, Yale book of numbers. Historical statistics of the college and university. New Haven, Yale Office of Institutional Research, pp 1701–1976
2. Postman N (1992) Technopoly the surrender of culture to technology. Alfred A. Knopf, New York
3. Kupiainen E, Mantylaa M, Itkonen J (2015) Using metrics in agile and lean software development—A systematic literature review of industrial studies
4. Unterkalmsteiner M, Gorschek T, Moinul Islam, AKM (2011) Evaluation and measurement of software process improvement—a systematic literature review
5. Dalcherand D, Benediktsson O, Thorbergsson H (2005) Development life cycle management: a multiproject experiment
6. Metrics in Agile Project Courses. In: 2016 IEEE/ACM 38th IEEE international conference on software engineering Companion
7. Mkiaho P, Poranen T, Seppi A (2015) Software metrics in students' software development projects. In: Proceedings of international conference on computer systems and technologies (CompSysTech'15)
8. Atkinson R (1999) Project management: cost, time and quality, two best guesses and a phenomenon, its time to accept other success criteria. Int J Proj Manag 17(6):337–342
9. Cooke-Davies T (2002) The real success factors on projects. Int J Proj Manag 20(3):185–190
10. Ika L (2009) Project success as a topic in project management journals. Proj Manag J 40(4):6–19
11. Jugdev K, Müller R (2005) A retrospective look at our evolving under-standing of project success. Proj Manag J 36(4):19–31
12. Baccarini D (1999) The logical framework method for defining project success. Proj Manag J 30(4):25–32
13. Bannerman P (2008) Defining project success: a multilevel frame-work. In: Proceedings of PMI research conference
14. Markus M, Mao J (2004) Participation in development and implementation—updating an old, tired concept for today's IS contexts. J Assoc Inf Syst 5:514–544

15. Shenhar A, Dvir D (2007) Project management research—the challenge and opportunity. Proj Manag J 38(2):93–99
16. McLeod L, Doolin B, MacDonel B (2012) A perspective-based understanding of project success. Proj Manag J 43(5):68–86
17. Ralph P, Kelly P (2014) The dimensions of software engineering success
18. Buse R, Zimmermann T (2012) Information needs for software development analytics. In: Proceedings of 20th international conference on software engineering. IEEE Press, pp. 987–996
19. ISO/IEC25010 (2011) Systems and software engineering
20. Kitchenham B, Pfleeger S (1996) Software quality: the elusive target. IEEE software
21. Naboulsi Z (2017) Code metrics—cyclomatic complexit. MSDN ultimate visual studio tips and tricks blog. https://blogs.msdn.microsoft.com/zainnab/2011/05/17/code-metrics-cyclomatic-complexity
22. McCabe Associates (1999) Integrated quality as part of CS699 professional seminar in computer science
23. Rosenberg L, Hammer T (1998) Software metrics and reliability. NASA GSFC
24. Rosenberg L, Hammer T (1998) Metrics for quality assurance and risk assessment. In: Proceedings of 11th international software quality week, USA
25. Brechner E (2003) Things they would not teach me of in college: what microsoft developers learn later. In: ACM SIGPLAN conference on object-oriented programming, systems, languages, and applications
26. Exter M, Turnage N (2012) Exploring experienced professionals' reflections on computing education. ACM Trans Comput Educ (TOCE) 12(3)
27. Bijlsma D, Ferreira M, Luijten B, Visser J (2012) Faster issue resolution with higher technical quality of software. Softw Qual J 20(2):265–285
28. Baggen R, Correia J, Schill K, Visser J (2012) Standardized code quality benchmarking for improving software maintainability. Softw Qual J 20(2):287–307
29. Correia J, Kanellopoulos Y, Visser J (2010) A survey-based study of the mapping of system properties to ISO/IEC 9126 maintainability characteristics. In: Proceedings of 25th IEEE international conference on software maintenance (ICSM), pp 61–70, 2009. Relation to software maintainability, Master thesis, University of Amsterdam
30. Visser J (2016) Building maintainable software: ten guidelines for future-proof code. OReilly Media, Inc
31. Booch G (1991) Object oriented design: with applications
32. Bruegge B, Krusche S, Alperowitz L (2015) Software engineering project courses with industrial clients
33. Olszewska M, Heidenberg J, Weijola M (2016) Quantitatively measuring a large-scale agile transformation. J Syst Softw
34. Palacin-Silva M, Khakurel J, Happonen A (2017) Infusing design thinking into a software engineering capstone course. In: Proceedings of 30th IEEE conference on software engineering education and training (CSEE&T)
35. Unterkalmsteiner M, Gorschek T (2012) Evaluation and measurement of software process improvement—a systematic literature review. IEEE Trans Softw Eng
36. Ilieva S, Ivanov P, Stefanova E (2004) Analyses of an agile methodology implementation. In: Proceedings of 30th EUROMICRO conference
37. Albrecht A (1979) Measuring application development productivity. In: Proceedings of joint SHARE, GUIDE, and IBM application development symposium, pp 83–92
38. Ochodek M, Nawrocki J (2011) Simplifying effort estimation based on use case points. Inf Softw Technol
39. Abrahamsson P (2003) Extreme programming: first results from a controlled case study. In: Proceedings of 29th EUROMICRO conference
40. Middleton P, Taylor P (2007) Lean principles and techniques for improving the quality and productivity of software development projects: a case study. Int J Prod Qual Manag

41. Hoegl M, Gemuenden H (2001) Teamwork quality and the success of innovative projects: a theoretical concept and empirical evidence. Organ Sci 12
42. G. Melnik and F. Maurer, "A Cross-Program Investigation of Students Perceptions of Agile Methods", International Conference on Software Engineering, 2005
43. Clark C (2005) Evaluating student teams developing unique industry projects. In: Proceedings of 7th australasian conference on computer education
44. Casey-Campbell M, Martens M (2008) Sticking it all together: a critical assessment of the group cohesion-performance literature
45. Carron A, Brawley L (2000) Cohesion: conceptual and measurement issues. Small Group Res 31
46. Salas E, Grossman R (2015) Measuring team cohesion: observations from the science. Hum Factors 57
47. Sommerville I (2004) Software engineering. Addison-Wesley
48. Carron A, Brawley L (2002) G.E.Q. the group environment questionnaire test manual. Fit Inf Technol 1135, Inc
49. Wellington C, Briggs T (2015) Comparison of student experiences with plan-driven and agile methodologies. In: 35th ASEE/IEEE frontiers in education conference
50. van Boxmeer F, Verwijs C (2007) A direct measure of morale in the 1140 Netherlands armed forces morale survey: theoretical puzzle, empirical testing and validation. In: Proceedings of international military testing association symposium (IMTA)
51. van Boxmeer F, Verwijs C, Euwema M (2011) Assessing soldier's morale in a challenging environment
52. Verwijs C (2018) Agile teams: don't use happiness metrics, measure team morale
53. Begel A, Simon B (2008) Struggles of new college graduates in their first software development job. In: Proceedings of 39th SIGCSE technical symposium on computer science education

Light-Weight Congestion Control for the DCCP: Implementation in the Linux Kernel

Agnieszka Chodorek and Robert R. Chodorek

Abstract The DCCP complements the UDP with three TCP-friendly congestion control modes. The most important are the CCID 2 mode, which uses the TCP's congestion control mechanism, and the CCID 3 mode, which implements the TFRC congestion control. The CCID 3 is intended for general multimedia transmission, although the authors of the DCCP specification notes that it may caused problems in the case of high-bandwidth video transmissions. This paper presents a prototype implementation of the modified TFRC congestion control, designed for multimedia transmission, that uses the RTP linear throughput equation (instead of the TCP one, originally used by the TFRC). This solution was introduced by the Authors in their previous work. The prototype implementation in the Linux kernel includes both a new congestion control module and updates to the DCCP kernel API. The proposed solution causes the DCCP to be not fully TCP-friendly, but still remains TCP-tolerant and does not cause unnecessary degradation of competing TCP flows. As a result, this method of congestion control can be used wherever it is necessity to use a TCP-tolerant protocol and when the rate of the media stream cannot be limited in a way typical for the TCP.

1 Introduction

The Datagram Congestion Control Protocol (DCCP) [1] was designed as the potential successor to the User Datagram Protocol (UDP). The purpose of the work on the new protocol was to respond to the increased growth of long-lived or large UDP

A. Chodorek
Faculty of Electrical Engineering, Automatic Control and Computer Science,
Kielce University of Technology, Al. Tysiaclecia P.P. 7, 25-314 Kielce, Poland
e-mail: a.chodorek@tu.kielce.pl

R. R. Chodorek (✉)
Department of Telecommunications, AGH University of Science and Technology, Al.
Mickiewicza 30, 30-059 Krakow, Poland
e-mail: chodorek@agh.edu.pl

© Springer Nature Switzerland AG 2020
A. Poniszewska-Maranda et al. (eds.), *Data-Centric Business and Applications*,
Lecture Notes on Data Engineering and Communications Technologies 40,
https://doi.org/10.1007/978-3-030-34706-2_13

streams that appeared in the Internet at the very beginning of 21st century. This growth of non-congestion-controlled streams, generated by the stand-alone UDP or by the UDP co-operating with the Real-time Transport Protocol (RTP)[1] was mainly caused by the fast growing market for Voice over Internet Protocol (Voice over IP, VoIP) applications. Much larger amounts of uncontrolled data might be waiting to be sent, as it seemed then, by the Video over IP market already standing in the starting blocks. Due to the high volume of traffic per single video stream, non-congestion-controlled applications like video-telephony and multiplayer games might become applications capable of killing the global Internet.

The relatively high reluctance of the application developers at that time to introduce application-level congestion control to their works did not give much hope that the problem would be solved. As a result, the creators of the new protocol assumed that *"most application writers would use congestion control for long-lived unreliable flows if it were available in a standard, easy-to-use form"* [1], and decided that they would deliver this *"standard, easy-to-use form"* to application developers.

The authors of the DCCP specification presents their protocol as the Transmission Control Protocol (TCP) *"minus bytestream semantics and reliability"* [2] (let's add to this: and minus flow control) and as the UDP protocol *"plus congestion control, handshakes, and acknowledgments"* [2]. The strength of the DCCP consists in the combination of the UDP's simplicity with congestion control. Nowadays, three modes of the DCCP's congestion control have been specified by the Internet Engineering Task Force (IETF) as Request For Comments (RFC) documents. The first is a direct implementation of the TCP's congestion control mechanism [3], the second introduces to the DCCP the TCP-Friendly Rate Control (TFRC) [4], and the third is a modification of the TFRC intended for small packets [5] (for example, packets that convey voice samples). The DCCP incorporates both implicit congestion notification (classic congestion notification based on packet loses) and explicit congestion notification [6] (based on Explicit Congestion Notification, or ECN, bits set in the header of IP packets [7]).

Although the Datagram Congestion Control Protocol is mainly focused on the— *"nomen omen"*—congestion control, the protocol also tries to solve other problems afflicting UDP transmissions, such as Network Address Translators (NATs) and other middlebox traversals, and security issues. Initially the DCCP was designed as a stand-alone transport protocol, but later the specification of the protocol was updated with the ability to replace the UDP when it co-operates with the Real-time Transport Protocol (RTP).

The DCCP is relatively old protocol (its specification, RFC 4340, is dated March 2006), but it still remains an interesting subject of research. The DCCP was extensively tested, and during tests both in simulation [8–11] and in a real-world environment [12] some anomalies were observed, such as unexpectedly large error rates. Analysis of DCCP behaviour led to the proposition for both the enhancement of the existing CCID 3 congestion control mechanism [11], and changes to the implementation of the CCID 3 in the Linux kernel [12]. The proposition for one of the

---

[1]Either the UDP, or the RTP do not provide congestion controlled transmission.

most revolutionary changes to the DCCP was presented in the paper [13], where a light-weight reliability assurance was proposed.

The TFRC, which was adopted as the CCID 3 mechanism, was also tested in terms of the performance evaluation and the Quality of Service (QoS) of the multimedia traffic controlled by this protocol. Tests were carried out for the standard TFRC [14, 15], complying with the RFC 5348 [16], and for the modified TFRC that has changed the classic TCP throughput equation to the rate based Square Increase Multiplicative Decrease (rate based SIMD, RB-SIMD) [15]. The RB-SIMD congestion control was a main part of the RTP RB-SIMD (RRB-SIMD) protocol, which achieved less frame loss rate and a larger peak-to-noise ratio than the classic TFRC [17].

Another solution of the TFRC congestion control was presented in the paper [18], where the classic, nonlinear TCP throughput equation was replaced by the linear RTP throughput equation. This light-weight congestion control (the protocol's processing effort is much smaller than if the complex, standard equation was used) does not assure full TCP-friendliness, manifested by the fair share of network bandwidth [16] (understood as equality towards competing TCP flows). However, it assures TCP-tolerance (understood as the avoidance of unnecessary degradation of competing TCP flows [19]).

The modified TFRC [18] was later used in the DCCP, as a modified CCID 3 congestion control mechanism [20, 21]. This mechanism was tested in a simulation environment [18], and then in a mixed, real and emulated environment [21]. In experiments [21], the preliminary version of the implementation in the Linux kernel of the modified CCID 3 with a linear throughput equation was used.

In this paper we describe a mature version of the prototype implementation of this DCCP's congestion control in the Linux kernel. The rest of the paper is organized as follows. The second section presents the DCCP standard ecosystem. The third section describes the architecture of the DCCP protocol implementation in the Linux kernel, while the fourth section presents implementation issues, namely a new congestion control module and changes introduced to the DCCP kernel API. The last, fifth section summarizes our experiences.

2   DCCP Standards Suite

2.1   Ecosystem of DCCP Standards

The first documented work on the Datagram Congestion Control Protocol began in 2002, when a set of IETF working documents (Internet Drafts) were published. A few years later, after several updates, these documents were published in a mature form as the suite of proposed standards that specify a new transport protocol for real-time multimedia transmission: the DCCP.

The specification of the DCCP includes several RFCs of different statuses that make up the DCCP standard ecosystem (Fig. 1). They belong to the Standards Track category and make up the DCCP standard suite [2–4, 22–27], or they belong to the

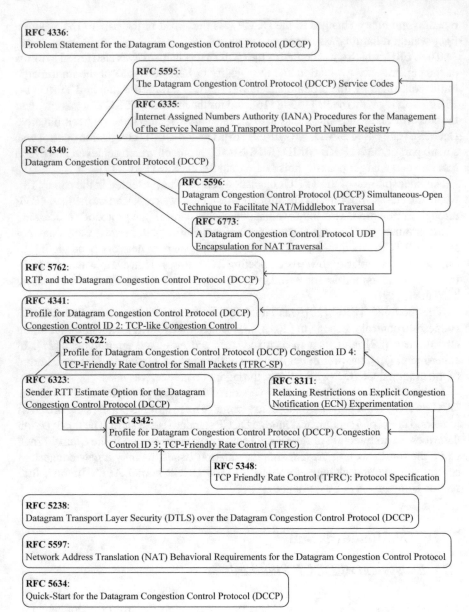

Fig. 1 Ecosystem of DCCP standards

Standards Track but they were not dedicated to the DCCP [6, 16], or they are not included in the Standards Track (categories: Informational [1], Best Current Practice [28, 29], Experimental [5, 30]), or they are not at the stage of an Internet Draft yet. A number of the RFC documents update other RFCs, which is symbolized in Fig. 1 by lines with arrows (an arrow shows a document that was updated).

Currently, the DCCP standards suite consists of 9 documents that are including in the category Standards Track. Their current status is Proposed Standards, which means they are at the entry level of maturity in the Standards Track [31]. Proposed Standards are, generally, stable and relatively mature (when compared to RFCs that are not a part of the Standard Track), and since they are still the subject of research, it is likely that there will be some changes to the documents in the more or less near future.

2.2 Datagram Congestion Control Protocol

The suite of the RFCs belonging to the Standards Track begins with a set of three successive documents (RFC 4340 [2], RFC 4341 [3], and RFC 4342 [4]), which specifies the DCCP transport protocol, and then updates the specification with two congestion control profiles. The first profile, the TCP-like Congestion Control (TCP CC), introduces the DCCP window-based congestion control that uses the classic Additive Increase Multiplicative Decrease (AIMD) algorithm. The second profile introduces the TFRC congestion control (TFRC CC) to the DCCP, which emulates the behaviour of the TCP protocol under congestion with the use of the so-called TCP throughput equation (the equation that describes the throughput of the TCP as a function of error rate).

Although the CC in DCCP stands for Congestion Control, the protocol does not apply any native congestion control mechanism (although it has native congestion notification, both implicit and explicit), and uses external CC profiles, standardized as separate RFC documents (Fig. 2). The use of external profiles instead of internal mechanisms causes that the DCCP provides open architecture that may be relatively easy updated with future CC profiles to meet new challenges and meet requirements of new applications.

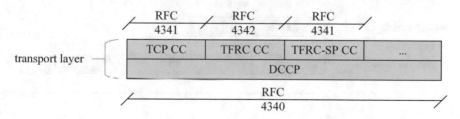

Fig. 2 DCCP and congestion control profiles

2.3 Security

RFCs from 4340 to 4342 are dated March 2006. Two years later, in May 2008, the
fourth RFC belonging to DCCP standards suite was published. It was the RFC 5238
[22], which describes the co-operation between the DCCP and the Datagram Transport Layer Security (DTLS) protocol. The DTLS [32] is a modification of the popular
Transport Layer Security (TLS) protocol. While the TLS protocol provides cryptographic protection of reliable data transmissions (usually it is used for cryptographic
protection of application data conveyed in TCP packets), the DTLS is oriented on
transmissions that do not assure reliability.

The RFC 5238 states that DTLS records are encapsulated in DCCP datagrams
in an N-to-1 manner (one DCCP packet must contain one or more complete DTLS
record). Both protocols are connection-oriented, which denotes that before the transmission of the application data both connections must be established. The sequence
of the handshakes is bottom-up, i.e. the DCCP connection is established first, and
then (using the established transport connection) the DTLS handshake procedure
is performed. The lifetime of the DTLS connection is limited from the top by the
lifetime of the underlying DCCP connection. There is no possibility of continuing
the DTLS connection if the underlying DCCP connection is closed.

Because the DTLS is placed in the presentation layer (Fig. 3), the protocol is
able to secure only data transmitted by the transport protocols, and it is not able to
assure the privacy of the transport protocols. If cryptographic protection of the DCCP
headers is needed, IP Security (IPSec) should be used.

2.4 Service Codes and NAT Traversal

In September 2009 the next two documents belonging to the Standards Track were
published, namely RFC 5595 [23] and RFC 5596 [24]. Both documents updated the
specification of the DCCP: the first with the description of how to use the 32-bit
Service Codes, the latter with the NAT/middlebox traversal.

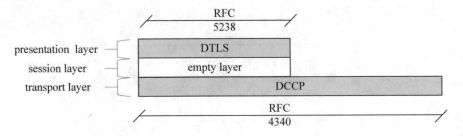

Fig. 3 DTLS over the DCCP

The Service Codes were defined in the RFC 4340 as identifiers of application-level services. The term application-level should be understood as the application program level or Process Level/Application layer of the Department of Defense (DoD) reference model rather than the application layer of the Open Systems Interconnection (OSI) Reference Model, because they should be assigned, for example, for the Session Initiation Protocol (SIP) control service (session layer of the OSI model) or the RTP media transmission service (transport layer of the OSI model).

The Service Code creates a 32-bit addressing space similar to the port addressing space. However, while Port Numbers are used for the multiplexing and demultiplexing of transport connections, Service Codes are used for the multiplexing and demultiplexing of application-level services over the transport connections. As a result, if two or more applications (servers of a given service), running on the same station and identified by the same IP address, want to use the same port numbers (listen on the same port number, assigned to the given service), they can be distinguished by the application-specific Service Code.

2.5 Protocol Stack

The DCCP provides transport services alone, without any co-operation with any transport protocols [2], or it can replace the UDP in the RTP/UDP protocol stack [25]—the most popular protocol stack for real-time multimedia transmission. In the case of stand-alone work, the DCCP occupies the whole transport layer of the Open Systems Interconnection (OSI) Reference Model, and in the case of co-operation with the RTP it occupies only the lower sublayer of the transport layer (Fig. 4). The cooperation of the DCCP and the RTP also applies to the Real-time Transport Control Protocol (RTCP), the signalling protocol associated with the RTP and used by the RTP to monitor the quality of the transmission.

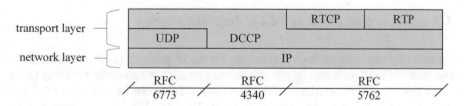

Fig. 4 DCCP/IP, DCCP/UDP/IP and RTP/DCCP/IP protocol stack

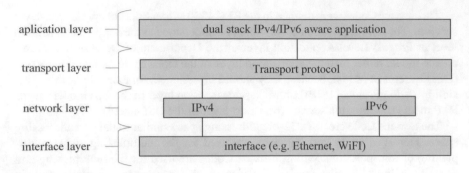

Fig. 5 Dual IP stack (IPv4 and IPv6)

3 The DCCP Implementation

Over the past few years, many implementations of the DCCP protocol have been developed. They have been implemented in the kernel space (for Linux and FreeBSD operating systems) or in the user space. However, almost all of the experimental DCCP implementations currently are unmaintained and abandoned.

Currently the commonly used DCCP implementation is implemented in the kernel space which is included in the official Linux kernel development tree (since version 2.6.14). This implementation was written by Arnaldo C. Melo. Some parts of the DCCP implementation share code with Linux's TCP implementation and also with other Linux implementations of INET[2] transport protocols, such as the Stream Control Transmission Protocol (SCTP).

The DCCP implementation uses the standard socket application programming interface (API) in the same way as the TCP does. Therefore, the code of user programs, written for server and client, is similar to the code of a regular TCP application.

3.1 The Architecture of DCCP Implementation

Current transport protocols works directly on the IP protocol (IPv4 or IPv6). Both versions of the IP protocols will continue to operate simultaneously for some time. In RFC 4213 [33] the methods of the transition mechanism from IPv4 to IPv6 are described. A dual-stack implementation of the IPv4 and IPv6 is the easiest way to do this transition. Dual-stack IP implementations provide complete IPv4 and IPv6 protocol stacks in the operating system (Fig. 5).

In the operating systems (including Linux), implementations of transport protocols are closely related to the IP protocol. Transport protocols do not require functional changes when cooperating with different versions of the IP protocol. However,

---

[2]A group of internet protocols in the Linux system.

implementations in operating systems require modifications to provide concurrent support for IPv4 and IPv6.

The DCCP implementation is working directly on top of the IP protocol (IPv4 or IPv6). Because the API for each IP version is different in the system, it is necessary to build two kernel DCCP modules: one for IPv4 (Fig. 6) and a second one for IPv6 (Fig. 7).

The DCCP implementation in the kernel space is included in the official Linux kernel development tree and its stored in the directory /net/dccp/. Depending on the initial configuration of the Linux kernel, the implementation can be loaded one of two ways: immediately (loaded by the system during the start of the Linux) or on demand (loaded by the user).

The DCCP implementation was divided into a base protocol described in [2] (stored in the directory /net/dccp/) and pluggable congestion control modules which implement desired congestion control strategies (stored in the directory /net/dccp/cids/).

3.2 Connection Management

The DCCP protocol is a connection-oriented transport protocol. It has connection management which has three phases: connection establishment, data transfer and connection termination.

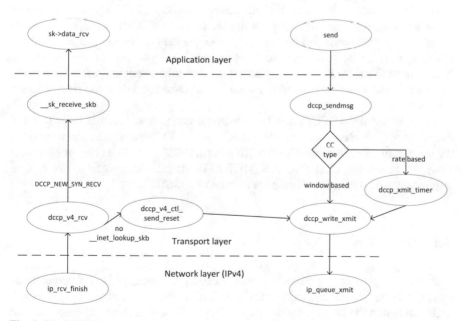

Fig. 6 The DCCP implementation working on top of the IPv4 protocol

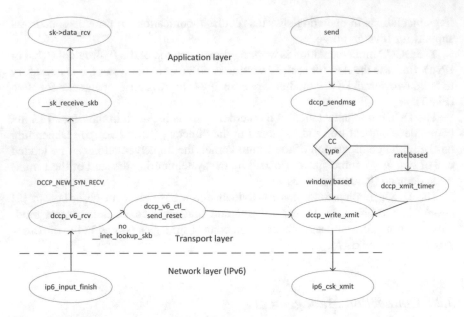

Fig. 7 The DCCP implementation working on top of the IPv6 protocol

During the lifetime of a DCCP connection the local end-point undergoes a series of state changes (Fig. 8).

The DCCP connection is established between two hosts. The first host, the initially passive host (often called the server) waits for an active DCCP connection ("passive open" in Fig. 8). The second host actively initiates the connection (often called the client) by sending the Request packet ("active open" in Fig. 8). A three way handshake is then used to set up a DCCP connection between the client and server. After the three way handshake the connection is now established and data can be sent (state OPEN in Fig. 8).

Termination of the DCCP connection requires a handshake between hosts with the exchange of CloseReq, Close and Reset packets (Fig. 8). Typically the server changes the state from OPEN through CLOSEREQ, to CLOSED. The client changes the state from OPEN through CLOSING to TIMEWAIT. After 2MSL wait time (in a typical implementation 4 min) the client goes to state CLOSED.

3.3 Data Flow

In the DCCP protocol connections are bidirectional. Internally DCCP connections consist of two separate unidirectional connections (called half-connections). For each half-connection the DCCP protocol allows the setup of different congestion control mechanisms and/or different mechanism parameters.

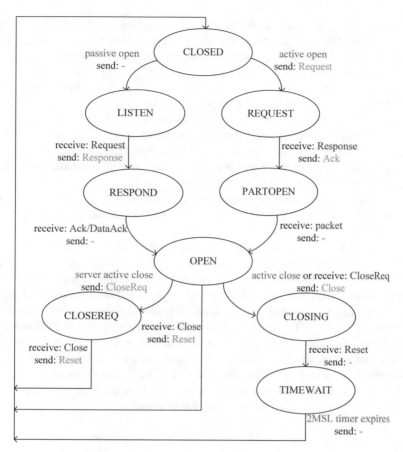

Fig. 8 The DCCP state diagram

Applications which use the DCCP protocol send and receive data using the socket interface (described in Sect. 3.4). The user data (payload) from the DCCP sockets, that were created in userspace, can be send by system calls `send()` handled in the kernel space by the `dccp_sendmsg()` method (defined in `/net/dccp/proto.c`) (Fig. 6 and Fig. 7). This data will be sent in the form of DCCP packets. Before it starts sending, the `dccp_sendmsg()` method, checks on whether the DCCP connection has already been established and on whether it is in the OPEN state. If no connection has been established yet, the system call waits in `sk_stream_wait_connect()` for a connection.

Next the data to be sent is processed depending on the type of congestion control (window-based or rate-based) chosen for the connection. For window based congestion control, packets to be send directly using the `dccp_write_xmit()` method (defined in the file `net/ddcp/output.c`). For rate based congestion control, packets to be send later via `dccps_xmit_timer()` expiry which calls the `dccp_write_xmit()` method (Figs. 6 and 7).

Before sending any packet from the DCCP, the output method `dccp_write_`
`xmit()` waits for the permission to send the packet from the congestion control
module (selected CCID module used for current direction of transmission). To do that
operation the output method calls the `ccid_hc_tx_send_packet()` method.
The return code of the `ccid_hc_tx_send_packet()` method determines if and
when the packet can be sent. It is possible to send the packet immediately, send it
after a specified time, leave it in the queue (from which it will be sent again later by
the same output method) or reject the packet (remove from the queue and not send
anymore). Finally, the data goes to the network layer specific `queue_xmit()`
sending callback. This callback uses the `ip_queue_xmit()` method for IPv4 and
the `inet6_csk_xmit()` method for IPv6.

The packet received from the network layer (from method `ip_rcv_finish()`
for IPv4 or `ipv6_input_finish()` for IPv6) goes to the `dccp_v4_rcv()`
method for IPv4 or `dccp_v6_rcv()` for IPv6.

Incoming packets from the network layer are analyzed - some sanity checks are
made (e.g. checking to see if the packet size is shorter than the DCCP header) and
invalid packets are discarded. Then the packets are classified and a lookup for a
corresponding socket is performed (a lookup in the established sockets hash table).
Depending on packet type and state of the DCCP protocol, the `dccp_v4_rcv()` (or
the `dccp_v6_rcv()` for IPv6) calls appropriate actions. If the connection is open
(state: OPEN) and data packets arrive, they are passed to the `_sk_receive_skb`
method. If data is received and the connection is not in the OPEN state the connection
reset signal will be sent by calling the `dccp_v4_ctl_send_reset()` method
for IPv4 or `dccp_v6_ctl_send_reset()` for IPv6. Reset is also called during
an incorrect sequence of packets initiating the connection.

3.4 Socket Interface

The DCCP protocol uses the socket interface[3] to send/receive data from/to applications. The sockets API uses several methods to: create a new socket, associates a
socket with a local port and an IP address, send and receives a message etc.

During the DCCP initialization, a `proto` object is defined (`dccp_v4_prot`
for IPv4 and `dccp_v6_prot` for IPv6) and its specific callbacks for DCCP are set
(Table 1). Table 1 shows two classes of methods. The first class of methods (containing the `dccp` name) is dedicated to the DCCP protocol and is used for the DCCP
specific operation. The second class of methods (containing the `inet` or `inet6`
name) is universal and can be used by various protocols such as TCP and are used,
among others, to search for the proper connection context (the appropriate socket
associated with the given transmission) or to initiate acceptance of the incoming
connection.

---

[3]The standard POSIX socket API which is used in Linux system is based on Berkeley sockets API
(BSD sockets).

Table 1 Socket callbacks for the DCCP

system call	callbacks in dccp_v4_prot	callbacks in dccp_v6_prot
.close	dccp_close	dccp_close
.connect	dccp_v4_connect	dccp_v6_connect
.disconnect	dccp_disconnect	dccp_disconnect
.ioctl	dccp_ioctl	dccp_ioct
.init	dccp_v4_init_sock	dccp_v6_init_sock
.setsockopt	dccp_setsockopt	dccp_setsockopt
.getsockopt	dccp_getsockopt	dccp_getsockopt
.sendmsg	dccp_sendmsg	dccp_sendmsg
.recvmsg	dccp_recvmsg	dccp_recvmsg
.backlog_rcv	dccp_v4_do_rcv	dccp_v6_do_rcv
.hash	inet_hash	inet6_hash
.unhash	inet_unhash	inet_unhash
.accept	inet_csk_accept	inet_csk_accept
.get_port	inet_csk_get_port	inet_csk_get_port
.shutdown	dccp_shutdown	dccp_shutdown
.destroy	dccp_v4_destroy_sock	dccp_v6_destroy_sock
.rsk_prot	&dccp_request_sock_ops	&dccp6_request_sock_ops
.orphan_count	&dccp_orphan_count	&dccp_orphan_count
.twsk_prot	&dccp_timewait_sock_ops	&dccp6_timewait_sock_ops

The DCCP protocol uses the socket interface in this same way as other connection-oriented transport protocols implemented in Linux operating system: TCP and Stream Control Transmission Protocol (SCTP). Data exchange between two applications (running on different hosts) using the DCCP protocol requires that the applications create sockets (in each host one socket) that will utilize the DCCP protocol (Fig. 9). A userspace application should create a DCCP socket by call the socket():

```
sock_fd = socket(domain, type, protocol);
```

The parameters of the socket() system call are:

- domain—communication domains that share common communication properties, such as naming conventions and protocol address formats—e.g. AF_INET for IPv4, AF_INET6 for IPv6,
- type—socket type—for the DCCP protocol it is the SOCK_DCCP argument,
- protocol—a valid IP protocol identifier—for DCCP is an IPPROTO_DCCP argument.

The return value of the socket() system call is the descriptor of this socket—an integer number that uniquely identifies the socket (it is used as a parameter to the next calls with this socket).

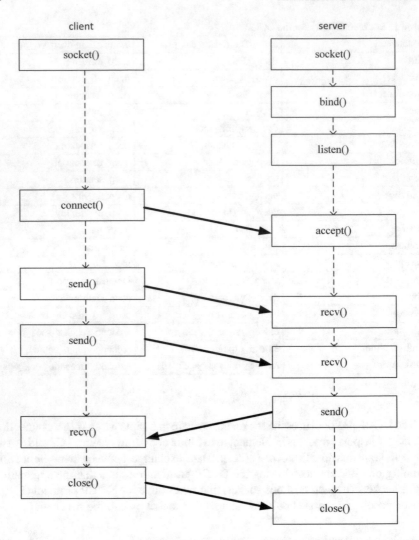

Fig. 9 Socket interface system call during sample client server data exchange using the DCCP protocol

Figure 9 shows the sequence of socket interface calls during the sample communication of two systems (client and server) which use the DCCP protocol.

At the beginning of the sequence (Fig. 9), a DCCP socket is created both in the client and server system using the socket() system call:

```
int dccpSocket = socket(PF_INET, SOCK_DCCP, IPPROTO_DCCP);
```

The server performs a "passive open" of the DCCP connection (Fig. 8). This must be done before the "active open" of the DCCP connection performed by the client.

After the creation of the socket, on the server side, the next step is to bind the socket to a specific port number by making the `bind()` system call. After binding, the next process is to request the kernel to start listening on the given port, which is done by making a call to `listen()`. Now the server is ready for connection.

The client "actively open" connection is done by making a call to `connect()`.

After the three way handshake the server performs the final step. It is a call to `accept()` systemcall. The connection between the client and server now is in an OPEN state. Both systems can exchange data using system calls `send()` and `recv()`.

The DCCP connection ends after the system call `close()`.

4 Prototype Implementation of the Congestion Control Module

This chapter presents the proposed concept of CC for the DCCP protocol and its implementation in the Linux kernel. In the chapter the principles of the usage of the proposed CC module in multimedia applications are also presented.

4.1 Current CC Implementations for DCCP in Linux Kernel

The DCCP implementation included in the official Linux kernel development tree supports only two congestion control methods which are identified by congestion control identifier (CCID): the CCID2 [3] and the CCID3 [4].

Additional CC implementations for DCCP, not included in the mainline Linux kernel, are in the DCCP test tree [34], which consists of updates and bug fixes to the Linux DCCP implementation. Currently in the DCCP test tree [34] we can find an implementation of CCID4 [5] and CCID_CUBIC (an implementation of TCP-Cubic mechanism for DCCP, in previous implementations called CCID5 and sometimes also called CCID-249) included.

The DCCP implementation uses pluggable congestion control modules. Those modules are stored in the directory `/net/dccp/cids/` and they use libraries stored in the directory `/net/dccp/cids/lib/`. The congestion control CCID 2 and 3 modes are implemented as separate modules (dccp_ccid2 and dccp_ccid3, respectively).

In the DCCP protocol, the currently used CCID mechanism is negotiated when establishing a connection (individually for each half-connection). It can be later renegotiated and changed to another one.

An application using the DCCP protocol can use the default CCID mechanism chosen in the system or indicate which CCID mechanism is expected by calling the DCCP socket options dedicated for this purpose.

Default CCID settings for the DCCP protocol in the Linux system are defined by the values in variables `rx_ccid` and `tx_ccid` (`/proc/sys/net/dccp/default/rx_ccid` and `/proc/sys/net/dccp/default/tx_ccid`). Possible values depends on what is currently registered in the system CCID modules (e.g. if CCID3 is registered in the system it is possible to define value 3 for both variables).

The application can select the appropriate CCID for a given transmission by calling the socket interface using the following socket options:

- DCCP_SOCKOPT_AVAILABLE_CCIDS—gives list of CCIDs supported by the endpoint,
- DCCP_SOCKOPT_CCID—sets both the transmitting (TX) and receiving (RX) CCIDs at the same time,
- DCCP_SOCKOPT_RX_CCID—read or sets the current CCID for the RX CCID,
- DCCP_SOCKOPT_TX_CCID—read or sets the current CCID for the TX CCID.

4.2 The Proposed Congestion Control Mechanism for the DCCP Protocol

The congestion control mechanism works at the sender and decides to how many packets the sender can send to the network. In the DCCP protocol, data packets and acknowledgements are send under the congestion control mechanism.

The proposed congestion control mechanism for the DCCP protocol is based on the modified TFRC building block described in [18], which joins elements of RTP behavior with equation-based congestion control and the rate control mechanism of the TFRC. The RTP flow itself is not able to react to congestion, so the only reaction is packet losses in the intermediate routers. Packet losses reduce the bit rate of the flow linearly [18]. As a result, in a network that is well-dimensioned for multimedia traffic, the RTP throughput equation depends only on the bit rate of the carried multimedia stream and the packet error rate [18]:

$$T(PER) = BR(1 - PER) \tag{1}$$

where: BR is a bit rate of multimedia stream, PER is a packet error rate.

Expanding Eq. (1) to a more general form we get [20]:

$$T(PER) = BR(a - b \cdot PER) \tag{2}$$

where: a is the scale factor, b is the slope factor.

To calculate the sending rate the proposed congestion control mechanism for the DCCP protocol uses the packet error rate and the bit rate of the multimedia stream. The packet error rate is defined as the number of lost packets or the ratio of the Explicit Congestion Notification (ECN) marked packets to the total number of packets sent. The bit rate of the multimedia stream is defined by the sending application and is typically equal to the target bit rate of the sending multimedia stream.

4.3 Creating New Congestion Control Module

Our implementation of the new DCCP's congestion control module (dccp_lin) has been created for the Linux 4.14.94 kernel.

First, a new CCID must be allocated. For the congestion control module presented in the paper the first, currently not used, experiential CCID (CCID = 250)[4] was allocated. A new CCID (symbolic identifier DCCPC_CCID_LIN) has been added in the CCID enum in the file /include/uapi/linux/dccp.h:

```
/* DCCP CCIDS */
enum {
    DCCPC_CCID2 = 2,
    DCCPC_CCID3 = 3,
    DCCPC_CCID4 = 4,
#define DCCPC_TESTING_MIN    248
#define DCCPC_TESTING_MAX    255
    DCCPC_CCID_ZERO = DCCPC_TESTING_MIN,
    DCCPC_CCID_CUBIC,
    DCCPC_CCID_LIN,
};
```

The basic data needed for the proper working of the dccp_lin module are stored in the ccidlin_hc_tx_sock structure. The structure contains, for example, the current sending rate (tx_x), receive rate (tx_x_recv), current loss event rate (tx_p), mean packet size expressed in bytes (tx_s), estimate of current round trip time (tx_rtt), and the target bit rate of the transmitted multimedia stream (tx_tbr).

To use the congestion control module one is required to define the interface described by a struct of ccid_operations. This interface consist of calls specific to functions for defined congestion control. Those calls are defined in the generic congestion control module for DCCP (file net/ddcp/ccid.c) which will activate the congestion control module identified by CCID. If the kernel has compiled with the option CONFIG_IP_DCCP_LIN (kernel option to compile proposed

---

[4]Since the RFC4340 [2] the experimental CCIDs must use ids greater than 248.

congestion control module), the struct `ccid_operations` for DCCPC_
CCID_LIN is included (file `net/dccp/ccid.h`):

```
#ifdef CONFIG_IP_DCCP_CCID_LIN
  extern struct ccid_operations ccid_lin_ops;
#endif
```

The struct `ccid_operations` includes:

- CCID ID—CCID identifier allocated for the congestion control module,
- CCID-specific initialization routine (called before CCID startup),
- CCID-specific cleanup routine (called before CCID destruction),
- packet receiving routine for receiver half-connection side,
- parsing option routine for CCID half-connection specific options,
- insert option routine for CCID half-connection specific options,
- feedback processing routine for the half-connection sender,
- sending packet routine for the sending part of the half-connection sender,
- diagnostic message information routine for half-connection receiversender,
- socket options routine specific to half-connection receiversender,

The routines mentioned above are implemented in the base part of the congestion control module (files `net/dccp/ccids/ccid_li.c` and `net/dccp/ccids/ccid_li.h`).

To implement the proposed congestion control module new library files (in directory `net/dccp/ccids/libs`) were defined. The file `net/dccp/ccids/libs/lin.c` consists mostly of linked initiation functions for both the sender and receiver parts of a connection. It also contains a linked function to the congestion control equation, which can be found in `net/dccp/ccids/libs/lin_equation.c`. In the listing below these linked functions can be seen:

```
u32 lin_calc_x(u16 s, u32 TBR, u32 p);
u32 lin_calc_x_reverse_lookup(u32 fvalue);

int lin_tx_packet_history_init(void);
void lin_tx_packet_history_exit(void);
int lin_rx_packet_history_init(void);
void lin_rx_packet_history_exit(void);
```

The `lin_calc_x` function returns the transmit rate in bytes/second and calls with input parameters `s`, `TBR` and `p` where: `s` is the packet size in bytes, `TBR` is the target bit rate of the multimedia stream, `p` is the loss event rate.

4.4 DCCP Features for Congestion Control

Two DCCP endpoints can negotiate several DCCP protocol connection attributes called DCCP features. In the DCCP protocol many properties are controlled by DCCP features including the mechanism for congestion control selected individually for each half-connection.

Negotiations are performed at the beginning of the transmission during the connection initiation handshake. They can also be performed at any time during whole DCCP transmission.

To use the congestion control mechanisms required by the DCCP features, during the connection setup a features negotiation is performed (features negotiation is also used for DCCP parameters negotiation like the AckRatio). In the implementation the `dccp_feat_is_valid_sp_val()` function (in the directory `/net/dccp/feat.c`) is used to check the valid feature ID received by the end system. Extensions were defined in the identification of the Congestion Control ID feature:

```
switch (feat_num) {
case DCCPF_CCID:
    return (val == DCCPC_CCID2 || val == DCCPC_CCID3 ||
            val == DCCPC_CCID_LIN);
case DCCPF_SHORT_SEQNOS:
```

The congestion control module usually required some mechanisms in the receiver and sender side. In the DCCP implementation the required mechanism is defined by a dependency table. For our congestion control implementation the table is defined by the code:

```
static const struct
  ccid_dependency ccidlin_dependencies[2]\cite{ch13rfc3} = {
  { /* Dependencies of the receiver-side CCIDLIN */
    {    /* Send Loss Event Rate */
      .dependent_feat   = DCCPF_SEND_LEV_RATE,
      .is_local = true,
      .is_mandatory = true,
      .val       = 1
    },
    {    /* NDP Count */
      .dependent_feat   = DCCPF_SEND_NDP_COUNT,
      .is_local = false,
      .is_mandatory = true,
      .val       = 1
    },
    { 0, 0, 0, 0 },
```

```
    },
    { /* Dependencies of the sender-side CCIDLIN */
      {    /* Send Loss Event Rate */
      .dependent_feat   = DCCPF_SEND_LEV_RATE,
      .is_local = false,
      .is_mandatory = true,
      .val      = 1
      },
      {    /* NDP counts */
      .dependent_feat   = DCCPF_SEND_NDP_COUNT,
      .is_local = true,
      .is_mandatory = false,
      .val      = 1
      },
      { 0, 0, 0, 0 }
    }
};
```

Our implementation required mechanisms for the detection of errors (DCCPF_
SEND_LEV_RATE) and for the detection of the length of each burst of the non-
data NDP Count packets (DCCPF_SEND_NDP_COUNT) to be activated in both
sides (sender and receiver). The first mechanism is mandatory for both sides (option
is_mandatory is set to true). The second mechanism is mandatory only for the
receiving side (option is_mandatory is set to true only for the receiver side).

4.5 Extension to the Socket Interface

The existing socket API of the Linux DCCP implementation was supplemented by
additional functionality that allows applications to both put information about Target
Bit Rate (TBR) of the multimedia traffic and to get information about the packet and
bit error rate.

The first supplement allows applications to be properly setup for the RTP throughput equation. The setup of the throughput equation is made by the classic function
setsockopt(), which is called with a new parameter DCCP_SOCKOPT_TBR.
The DCCP implementation gets the delivered value of the TBR, which is interpreted
as an unsigned integer.

The second supplement allows applications to have easy co-operation with
the adaptive coder. Information about packet and bit error rate is determined by
the classic function getsockopt() called with new parameters, respectively,
DCCP_SOCKOPT_PER and DCCP_SOCKOPT_BER. The application gets information about packet and error rate in the format of a single precision floating-point
number.

In the header file (`net/dccp/ccids/ccid_li.h`) socket options for both the sender half and the receiver half of the connection are defined in structs.

5 Conclusions

In this paper the prototype implementation of the light-weight DCCP's congestion control algorithm in a Linux kernel was presented. The algorithm uses the RTP throughput equation, reported in a previous work of the Authors, and was designed for the purpose of multimedia transmission. The implementation consists of two parts. The first is a new congestion control module and the second is an improved DCCP kernel API. The prototype implementation has been validated and verified in a mixed (real and emulated) environment.

This work has also resulted in a much improved implementation of DCCP in the Linux kernel. The several not currently supported the features that are specified in RFCs 4340, 4341, 4342 was added. Also the contribution of several patches that fixed a variety of bugs and performance issues was noted.

Acknowledgements This work was supported by the Polish Ministry of Science and Higher Education with the subvention funds of the Faculty of Computer Science, Electronics and Telecommunications of AGH University.

References

1. Floyd S, Handley M, Kohler E (2006) Problem statement for the datagram congestion control protocol (DCCP). RFC 4336. https://doi.org/10.17487/RFC4336
2. Kohler E, Handley M, Floyd S (2006) Datagram congestion control protocol (DCCP). RFC 4340. https://doi.org/10.17487/RFC4340
3. Floyd S, Kohler E (2006) Profile for datagram congestion control protocol (DCCP) congestion control ID 2: TCP-like congestion control. RFC 4341. https://doi.org/10.17487/RFC4341
4. Floyd S, Kohler E, Padhye J (2006) Profile for datagram congestion control protocol (DCCP) congestion control ID 3: TCP-friendly rate control (TFRC). RFC 4342. https://doi.org/10.17487/RFC4342
5. Floyd S, Kohler E (2009) Profile for datagram congestion control protocol (DCCP) congestion ID 4: TCP-friendly rate control for small packets (TFRC-SP). RFC 5622. https://doi.org/10.17487/RFC5622
6. Black D (2018) Relaxing restrictions on explicit congestion notification (ECN) experimentation. RFC 8311. https://doi.org/10.17487/RFC8311
7. Ramakrishnan K, Floyd S, Black D (2001) The addition of explicit congestion notification (ECN) to IP. RFC 3168. https://doi.org/10.17487/RFC3168
8. Nor SA, Alubady R, Kamil WA (2017) Simulated performance of TCP, SCTP, DCCP and UDP protocols over 4G network. Proc Comput Sci 111:2–7. https://doi.org/10.1016/j.procs.2017.06.002
9. Rajput R, Singh G (2018) NS-2-based analysis of stream control and datagram congestion control with traditional transmission control protocol. In: Lobiyal D, Mansotra V, Singh U (eds) Next-generation networks. Advances in intelligent systems and computing, vol 638. Springer, Singapore. https://doi.org/10.1007/978-981-10-6005-2_31

10. Kokkonis G, Psannis KE, Roumeliotis M, Nicopolitidis P, Ishibashi Y (2017) Performance evaluation of transport protocols for real-time supermedia—HEVC streams over the internet. In: Proceedings of 12-th IEEE international symposium on broadband multimedia systems and broadcasting (BMSB), pp 1–5. https://doi.org/10.1109/BMSB.2017.7986185

11. Ibrahim MZ, Ibrahim DM, Sarhan A (2015) Enhancing DCCP-TCP-like mechanism for wireless sensor networks. In: Proceedings of tenth international conference on computer engineering & systems (ICCES), pp 234-240. https://doi.org/10.1109/ICCES.2015.7393052

12. Schier M, Welzl M (2012) Using DCCP: issues and improvements. In: Proceedings of the 2012 20th IEEE international conference on network protocols (ICNP), pp 1–9. https://doi.org/10.1109/ICNP.2012.6459967

13. Ahmed AA, Ali W (2018) A lightweight reliability mechanism proposed for datagram congestion control protocol over wireless multimedia sensor networks. Trans Emerg Telecommun Technol 30:3. https://doi.org/10.1002/ett.3296

14. Wheeb AH, Morad AH, AL Tameemi MI (2018) Performance evaluation of transport protocols for mobile ad hoc networks. J Eng Appl Sci 13(13):5181–5185. https://doi.org/10.3923/jeasci.2018.5181.5185

15. Priyanka E, Reddy PCh (2016) Performance evaluation of video streaming using TFRC, modified TFRC over wired networks. In: Proceedings of international conference on communication and electronics systems (ICCES). https://doi.org/10.1109/CESYS.2016.7889938

16. Floyd S. Handley M, Padhye J, Widmer J (2008) TCP friendly rate control (TFRC): protocol specification. RFC 5348. https://doi.org/10.17487/RFC5348

17. Lusilao Zodi GA, Ankome T, Mateus J, Iiyambo L, Silaa J (2019) A Unicast rate-based protocol for video streaming applications over the internet. In: Barolli L, Xhafa F, Khan Z, Odhabi H (eds) Advances in internet, data and web technologies. EIDWT 2019. Lecture Notes on Data Engineering and Communications Technologies, vol 29. Springer, Cham. https://doi.org/10.1007/978-3-030-12839-5

18. Chodorek A, Chodorek RR (2010) Streaming video over TFRC with linear throughput equation. Adv Electron Telecommun 1(2):26–29

19. Chodorek A, Chodorek RR (2009) An analysis of TCP-tolerant real-time multimedia distribution in heterogeneous networks. Traffic and performance engineering for heterogeneous networks. River Publishers, Wharton, TX, USA, pp 315–336

20. Chodorek R. Chodorek A (2014) Szybki mechanizm przeciwdzialania przeciazeniom dla protokolu DCCP realizujacego transmisje czasu rzeczywistego. Przeglad Telekomunikacyjny, Wiadomosci Telekomunikacyjne 87:835–841 (in Polish)

21. Chodorek RR, Chodorek A (2018) Light-weight congestion control for the DCCP protocol for real-time multimedia communication. Computer networks. In: International conference on computer networks. Springer, Cham, pp 52–63. https://doi.org/10.1007/978-3-319-92459-5_5

22. Phelan T (2008) Datagram transport layer security (DTLS) over the datagram congestion control protocol (DCCP). RFC 5238. https://doi.org/10.17487/RFC5238

23. Fairhurst G (2009) The datagram congestion control protocol (DCCP) service codes. RFC 5595. https://doi.org/10.17487/RFC5595

24. Fairhurst G (2009) Datagram congestion control protocol (DCCP) simultaneous-open technique to facilitate NAT/Middlebox traversal. RFC 5596. https://doi.org/10.17487/RFC5596

25. Perkins C (2010) RTP and the datagram congestion control protocol (DCCP). RFC 5762. https://doi.org/10.17487/RFC5762

26. Renker G, Fairhurst G (2011) Sender RTT estimate option for the datagram congestion control protocol (DCCP). RFC 6323. https://doi.org/10.17487/RFC6323

27. Phelan T, Fairhurst G, Perkins C (2012) A datagram congestion control protocol UDP encapsulation for NAT traversal. RFC 6773. https://doi.org/10.17487/RFC6773

28. Denis-Courmont R (2009) Network address translation (NAT) behavioral requirements for the datagram congestion control protocol. RFC 5597, BCP 0150. https://doi.org/10.17487/RFC5597

29. Cotton M, Eggert L, Touch J, Westerlund M, Cheshire S (2011) Internet assigned numbers authority (IANA) procedures for the management of the service name and transport protocol port number registry. RFC 6335, BCP 0165. https://doi.org/10.17487/RFC6335

30. Fairhurst G, Sathiaseelan A (2009) Quick-start for the datagram congestion control protocol (DCCP). RFC 5634. https://doi.org/10.17487/RFC5634
31. Bradner S (1996) The internet standards process—revision 3. RFC 2026, BCP 0009. https://doi.org/10.17487/RFC2026
32. Rescorla E, Modadugu N (2006) Datagram transport layer security. RFC 4347. https://doi.org/10.17487/RFC4347
33. Nordmark E, Gilligan R (2005) Basic transition mechanisms for IPv6 hosts and routers. RFC 4213. https://doi.org/10.17487/RFC4213
34. DCCP Test Tree. https://github.com/uoaerg/linux-dccp. Accessed June 2019